Project **Management**

To Madeline Healy (nee Winters) my mother
and in memory of Patrick Healy, my father

Project **Management**

Getting the job done
on time
and in budget

Patrick L. Healy

Australia
Butterworth-Heinemann, 18 Salmon Street,
Port Melbourne, Vic. 3207
Singapore
Butterworth-Heinemann Asia
United Kingdom
Butterworth-Heinemann Ltd, Oxford
USA
Butterworth-Heinemann, Newton

National Library of Australia Cataloguing-in-Publication data:

Healy, Patrick L., 1945–
Project management: getting the job done on time and in budget

Includes index
ISBN 0 7506 8943 9

1. Industrial project management. 2. Construction management. I. Title

658.404

Enquiries should be addressed to the publisher
Typeset in Australia by Diana Piper
Printed in Singapore by Chung Printers

Foreword

When Patrick Healy asked me to write a foreword for this book, my first question was, 'What does one write about in a foreword?'. The purpose of a foreword is to introduce the book, and the task of the writer of forewords is evidently to do this in a way which is relevant to both the book and the writer's own experience. Patrick knew of my long-time interest in the history of the development of modern project management and that in some areas I have been part of that development. He therefore suggested that I give a personal historical perspective on the development of project management in my time, and introduce this book in that context. This is what I have tried to do in the following.

When I was introduced to the practice of 'project management' some 50 years ago, it was not recognised or described as such. With hindsight, this may seem somewhat surprising, as projects are older than the pyramids. But modern recognition of *project management* as a distinctive practice appears to have had its genesis in the 1950s.

In North America, Bechtel's 1951–53 Transmountain Oil Pipeline in Canada appears to have been one of the first projects in which that organisation functioned as the project manager and assigned an individual with undivided responsibility for integrating the whole project. In the same period, 'joint project offices' were being established by the US Air Force as an organisational mechanism for achieving integration in its aircraft production projects.

In Australia, Civil & Civic (a company which I subsequently joined in 1961) first used the project management approach in 1954–55 in a 'rescue mission' which focused on managing the feasibility and design of a major subdivision project in Sydney, converting it from a marginal investment to a successful venture. This approach (via a 'project engineer') was adopted for all the company's subsequent property development projects, although it was not until 1962 that the term 'project management service' was officially coined to describe that company's integrated approach to design and construction in the building and construction industries.

My first really significant involvement with project management techniques, as was the case for most of my generation, was with network scheduling (CPM, PERT) in the early 1960s. Indeed, in that decade, network scheduling was regarded by many people as being synonymous with project management—which of course was simply not the case. But most of the books in that decade were about network scheduling, and it wasn't until the publication of Cleland and King's *Systems Analysis and Project Management* (McGraw-Hill) in 1968 that we got a much broader treatment of what project management was about. This was a particularly important book for me, because it reached me at a time when I was becoming increasingly aware that many of the conventional wisdoms of general management, which I had been teaching in-company for some time, were not appropriate for the project environment in which I was working. Thus began a search for an underlying theory and/or framework for project management, which has continued to this day.

The increasing interest in project management in the 1960s was also reflected in the formation of Europe's International Project Management Association (IPMA), formerly known as INTERNET, which had its first meeting in 1965; and of North America's Project Management Institute (PMI) in 1969. It should also be noted that, in the 1960s, project management was mainly practised in the construction, defence and aerospace industries. But this changed dramatically during the next decade. By the end of the 1970s, Harold Kerzner, in the first edition of his book *Project Management: A Systems Approach to Planning, Scheduling and Controlling* (Van Nostrand Reinhold, 1979), was able to report that the concept behind project management had spread to virtually all industries. This book ranged over a very wide spectrum of project management and, as far as I was concerned, was the second 'classic' book on project management.

The 1970s saw the emergence and/or refinement of a wide range of tools and techniques, including WBS, OBS, responsibility assignment matrices, and 'earned value' methods; the investigation and adoption of various organisational forms to undertake projects (particularly matrix forms); and an increasing recognition of the distinctive nature of project management as an avocation/profession.

The increasing level of interest in Australia was reflected in the formation of the Project Managers Forum, the predecessor of the Australian Institute of Project Management, which had its first public meeting in Sydney in 1976. At that meeting, I spoke to a substantial audience on 'A Participative Approach to Construction Planning and Control', which reflected the background and interests of the majority of the audience. Expansion of project management into other industries was lagging behind North America at the time.

The 1980s saw increased efforts to represent project management as a structured discipline and approach. This had been my own interest for many years, but I had not succeeded in developing a satisfactory framework myself, and up to the mid-1980s had not seen any other serious efforts to do so. Therefore I was particularly attracted by the two versions of a document entitled the Project Management Body of Knowledge (PMBOK), which were published by the USA's Project Management Institute (PMI) in 1986 and 1987. These documents described project management in terms of its 'functions'. The later version specifically embraced the management of project scope, quality, risk, human resources, communications and contract/procurement, in addition to the 'traditional' time and cost management functions. The body of the PMBOK document thus consisted of eight 'project management functions', which were the primary descriptors of the body of knowledge in project management.

The PMBOK document itself had many inconsistencies, which I discussed in a number of articles, which in turn provided the impetus for PMI to appoint me in 1989 as Chairman/Director of their Standards Committee, whose principal task was to further develop the PMBOK document. Such a major task proved to be beyond my ability to manage effectively from Australia, and after two years I handed over the leadership to my deputy, Bill Duncan, whose team produced the current *A Guide to the Project Management Body of Knowledge* in 1996. This document still follows the functional approach, but is a much more complete and integrated document.

Meantime, I had been recruited by the University of Technology, Sydney (UTS), to develop a Master of Project Management (MPM) degree course, as from 1 January 1988. My previous work on project management in general, and on the PMBOK in particular, had led me strongly to the conviction that describing an avocation by its 'functions', or by its tools and techniques, as did most project management books to that time, was not the most useful way to help people who essentially wanted an answer to the question, 'I have just been given this project to manage. What steps do I now need to take to effectively progress the project, from start to finish, to achieve a successful outcome?' I therefore decided to base the course on a backbone of what we called the 'project process', which in more conventional terms would be described as the 'project life cycle'. We developed a series of 'generic' processes for this, ie, processes which would be applicable to most projects in most application areas. Discussion of the project management 'functions', and tools and techniques, are introduced at appropriate places in the project process. Patrick Healy was involved in the development and delivery of this course and has followed a parallel approach in this book. But it should be emphasised that this book is a distinctively individual contribution and does not, and is not intended to, describe the composition of the UTS MPM course.

Essentially, this is a book for practising project managers and their teams, general managers involved with projects, project clients and other key stakeholders. This book augments the now very considerable material in the project manager literature. The author has not attempted to duplicate the well-known and well-documented 'functions' and tools and techniques of project management in the literature, but discusses their use in the context of the project life cycle.

He focuses on the practice of project management, and the book is replete with lists, check-lists,tests and the like, which should be useful for practitioners at all levels of experience.

Patrick Healy has made several innovative contributions in this book. Perhaps the most significant of these is an expanded project life cycle. Healy has added two phases to the four phases which are most commonly used as the primary breakdown of a generic project life cycle, namely Feasibility, Planning and Design, Implementation and Handover.

The first additional phase, which Healy calls Transition, covers the activities which are preliminary to the Feasibility phase. I regard this as a very valuable addition, particularly for general managers and project clients who are involved with possibilities of a project being initiated. As the author says, '...projects just don't come out of nowhere'. At some length, he discusses key issues, key players and factors such as agreement and support for the project, and concludes with a check-list of inputs, processes and outputs for the Transition Phase. The second additional phase nominated by Healy is Pre-Implementation. It could be argued that this might be incorporated into the previous and/or following phases. However, Healy uses it to discuss the many issues which need to be considered and resolved before the normally frenetic Implementation Phase gets under way. His check-lists relating to these issues are another very useful contribution.

Other distinctive features of his treatment of the project life cycle include a very substantial discussion of the management of the work in the Planning and Design phase and in the Implementation phase. In the latter, Healy discusses the responsibilities and management work of the client and of the main supplier separately, and discusses the functions of each as they apply both internally and externally to the project. In this way, he offers a multidimensional perspective on project implementation which is lacking in most other treatments.

Additional features of this book which I think are particularly useful include a chapter on how to enhance project sequence capability, an insightful discussion of interface management and a welcome chapter on management of the project from the client's point of view.

Although this is not an academic book, Healy has cited sources extensively in his appended notes to facilitate access to more detailed coverage of many of the topics discussed. For readers who are so interested, this also gives an enhanced appreciation of how the author has blended material from these sources with his own insights to produce a practical and pragmatic addition to the project management literature.

Alan Stretton
September 1997
Sydney, Australia

Contents

		Foreword	v
		List of Abbreviations	xii
		Introduction	1
PART ONE		OVERVIEW	
Chapter	**1**	**Project Management and its Social Impact**	9
	2	**The Overall Management of Project Acquisition**	31
PART TWO		THE PROJECT LIFE CYCLE	
Chapter	**3**	**Transition Phase**	55
	4	**Feasibility Phase**	73
	5	**Planning and Design Phase**	115
	6	**Pre-implementation Phase**	146
	7	**Implementation Phase**	169
	8	**Handover Phase**	194
	9	**Project Sequence Capability**	206
PART THREE		KEEPING IT ALL TOGETHER	
Chapter	**10**	**Project Communication**	227
	11	**Objectives and Scope Management**	250
	12	**Interfaces**	267
	13	**And finally the Client**	279
		Notes	285
		Further Reading	295
		Index	297

Acknowledgments

Getting this book through the early stages relied heavily on Helena Klijn, at that time my commissioning editor. I am most grateful to her for her support and guidance as the book took its main shape. The finishing tasks fell to Rosemary Peers, a person who knows how to manage to a deadline, who took over as commissioning editor, and to the editor, Liz Goodman, who took on the daunting job of cleaning up the written expression and making things shipshape.

I must also acknowledge the contribution (at the time unwelcome) made by a reviewer, who caused me to completely reorganise the text and end up with a better book.

Lynne Davis, our son David and daughter Tessa have tolerated my neglect of the important principle that life must still go on while the project sequence is being managed. Writing this book has been possible because they allowed me the freedom for this to happen.

Comments on sources of ideas

Most of the ideas in this book have turned up in other places and are the stuff of a range of disciplines. So the ownership of the ideas in this book is quite diverse. What I have attempted to do is assemble the ideas in a way that provides a holistic view of the management in project management, taking as a core the well-known idea called the 'project life cycle'.

I would like to acknowledge a debt to Adjunct Professor Alan Stretton for his overall perspective on project management and how it impacts on management in our society. I would also like to acknowledge the contribution to my ideas, in the late 1980s and early 1990s, made by (as he then was) Associate Professor Hamish MacLeannan of the University of Technology, Sydney.

In order to draw attention to some of the sources and in order to draw attention to the fact that others are also thinking about these ideas, some notes are provided at the end of the text. These provide pointers to other descriptions of the ideas or to other people who are thinking about similar issues. These notes are there to emphasise the wide arena from which project management draws its ideas and to introduce some ongoing thinking; they are not there to provide a rigorous sourcing of ideas. For the interested reader, probably the most complete exposition on the sources and development of ideas in project management is the book by P. W. G. Morris, *The Management of Projects.*

To any reader who believes that their contribution has been used and not acknowledged, the author apologises and asks that data be provided so that the error can be corrected in later editions.

Abbreviations

ACWP	Actual Cost of Work Performed
BATNA	Best Alternative to a Negotiated Agreement
BCWP	Budgeted Cost of Work Performed
BCWS	Budgeted Cost of Work Scheduled
CAD	Computer Aided Drafting
CPM	Critical Path Method
DCF	Discounted Cash Flow
EDI	Electronic Data Interchange
EIS	Environmental Impact Statement
IR	Industrial Relations
IRR	Internal Rate of Return
ISO	International Standards Organisation
IT	Information Technology
NLP	Neuro-Linguistic Programming
NPV	Nett Present Value
OBS	Organisation Breakdown Structure
OPIC	Objectives, Plan, Implement, Control
PBS	Product Breakdown Structure
PCG	Project Control Group
PERT	Program Evaluation and Review Technique
PLCA	Plan, Do, Check, Act
PMI	Project Management Institute
POIPH	Product of Implementation Phase
TQM	Total Quality Management
WBS	Work Breakdown Structure

Introduction

A project can be carried out in just about any sphere of human endeavour. It can involve massive space exploration, a huge million-line software program for submarines or making improvements to an ongoing work activity. Buying one's home or finding a home to rent is a project. Although cleaning one's house regularly is not a project, getting yourself organised to do the housework, or improving the way in which it is done, is a project. Typically, all projects reach a conclusion, that is, they have an end-point. Any activity with an end-point is a project; a repeated activity is not.

The need to establish some form of project management is not new. All the wonderful buildings of the ancient world—the vast structures the Egyptians built, the fine architectural achievements of the Greeks, the engineering works of the Romans, the medieval cathedrals of Europe, the Great Wall of China—all required project management. Battles that have had a decisive impact on the history of the world required project management: one side's project ended in success while the other side's project ended in failure. Allied to this, there has been a long history of developing weapons for war. Weapons development continues today and is a major consumer of project management skills. Various defence departments around the world have grappled with project management and have in the process undertaken much research which has brought a degree of standardisation to the area.

While project management as an activity is not new, what *is* new is the recognition of project management as a discipline. It is seen by many as a newly invented, special form of management, originating somewhere in the 1940s or 1950s; some believe it really started with the invention of PERT (Program Evaluation and Review Technique).

Its exact origin is not of interest here, except to note that it has really only come to the fore in the second half of the twentieth century.[1]

Project management has been identified as being separate from ordinary or general management; the usual perception is that one either practises ordinary general management or one practises project management. This book takes a different view, expressed as follows:

Managers need to practise the skills of both general and project management, and the separation of the two areas of management is no longer necessary or desirable.

The practice of project management

It is difficult to point to any one theory of project management. Various techniques are associated with it and many people think that these constitute project management theory. Although these techniques are important elements that underpin project management practice, they are only part of it. This book refers to these techniques only to point out how they might be applied or how one might approach using them.

The following chapters promote the view that project management practice involves achieving objectives using the project life cycle framework. One progresses through the project life cycle using ideas based on the project life cycle itself and also using general management and project management methods, techniques and ideas. One picks the methods, techniques and ideas as required to suit the situation.

This book also makes the point that social and political agendas surround the activity of project management, and any practicing project manager needs to be aware of this. Since the techniques of project management are well understood and are fairly easy to apply, they have lost much of their power to maintain the project manager's political control. It is absolutely essential that the project manager keeps a clear view of the external environment, keeps sounding out the level and extent of support for the project, and develops sensitivity to any weakening in the political support for it.

Project management requires flexibility. One needs to have a clear idea of where one is going but be prepared to take different routes if necessary. A project manager trained in both general and project management is better equipped to deal with situations on a contingency basis and has the flexibility to adopt the appropriate management technique. When getting a project under way, the project manager should first list the steps—and then use them as no more than a checklist. To follow a set formula can put much at risk and distracts the project manager from responding to an ever-changing environment.

Many organisations have produced project management manuals, vast documents of great detail with many, many pages. It is far more important that project managers develop their own attitudes or approaches to problems that allow them to identify and respond to the critical issues. There will be enough paperwork on the project itself without adding project manuals to the clutter.[2]

Project management techniques

The better-known techniques are Work Breakdown Structure (WBS), Organisation Breakdown Structure (OBS), Product Breakdown Structure (PBS), Task Responsibility Matrix, Earned Value Techniques, Network Techniques, Configuration Management and Risk Management.[3]

Breakdown structures

A breakdown structure is a hierarchical arrangement of work or other item; its structure resembles an inverted tree. An ordinary organisational chart is a breakdown structure, showing how the parts of an organisation relate to each other. Three particular breakdown structures are used in project management, WBS, OBS and PBS, the most common being WBS. WBS originated in the United States and now has world-wide recognition; the Germans refer to it as Projektstrukturplan (PSP).[4, 5]

WBS identifies all the work to be done on the project and breaks it down into manageable packages or pieces of work. It arranges these into a hierarchical relationship which is useful in managing the scope of a project and in interface management. OBS shows which organisation is responsible for what part of the work—to avoid problems, it is important that OBS be compatible with WBS. PBS is the breakdown of a product through function down to component. It is a hierarchical structure and describes all the parts that have to come together to make the product.

Considerable creative effort is required to construct a breakdown structure. One has to define the elements or parts that are to be brought into a relationship. There must be discretion in their definition: these parts have to be separable, that is, they must be discrete, which in complex projects can be quite difficult. There also has to be compatibility between breakdown structures.[6]

Task responsibility matrix

The task responsibility matrix is a rectangular grid with the tasks to be done written down one side and the names of people or organisations written across the top. The level of responsibility someone has for a task is recorded at the intersection of their name and the task. Various codes can be used to indicate the levels of responsibility.

Earned value techniques

Cost control is exercised using budgets. The question arises as to how one measures the value of work done to date and how one evaluates that measurement. Earned value techniques involve measuring the actual cost of work performed (ACWP), the budgeted cost of work performed (BCWP), and the budgeted cost of work scheduled (BCWS). The results can then be presented as a graph and link cost and time variations. There are proprietary systems that perform these analyses.[7]

Network techniques

These are the hallmarks of projects—useful and vital techniques for their management. As time goes by, the development of cheaper and more powerful computer systems will make these techniques readily available. Network techniques can also be used to balance resources used on the project; the network highlights points of peak demand. The network also allows the project manager to identify the project cash flow. Techniques include CPM and PERT.[8]

Configuration management

The purpose of configuration management is to ensure compatibility across the project. It is an important technique for larger projects; on smaller projects, a full-scale configuration management process is probably not required. The process involves using baselines to control the development of the definition of the scope and content of parts of the project. Terminology includes 'functional baseline', 'allocated baseline' and 'product baseline'; 'configuration accounting', 'control' and 'auditing'. Configuration management follows set procedures for the approval of variations, which includes the appointment of a configuration committee.[9]

Risk management

Since projects can be risky undertakings, risk management has recently become the focus of much research. It involves identifying the risks and deciding how they will be dealt with. Dealing with the risk could involve assuming liability for the risk, insuring against the risk, or transferring the risk to somebody else (which may include insuring the risk).[10, 11]

Generic project management

A question that interests many is whether or not there is such a thing as generic project management. Is there, in fact, a form of project management that can be applied across all industries? To some extent, there is—one finds the application of cost management to be very similar across the board, as is time management.

Within any one industry, it is almost certain that there is a generic form of management appropriate for that industry. There will a wide range of tasks that are common from project to project. In the project management of a building, many problems and issues will be met that were encountered with previous buildings and which will require the same approach. In the management of the introduction of one pharmaceutical drug, many issues and problems will arise that will be encountered during the introduction of the next one. Political campaigns have many common features, irrespective of political persuasion; any seasoned politician can tell you that the management steps vary little between campaigns.

However, when one jumps across industries, one finds that while much is common there are specific differences. The pharmaceutical industry has quite specific life cycle requirements that are different to those of the building industry, for example. Becoming involved in activities that cross industries calls for generic or general project management skills combined with some industry-specific project management skills. Many project managers can develop these industry-specific skills as the project proceeds; those working in an industry that is new to them should talk to people in that industry in order to identify the nature of the industry-specific tasks. It is hoped that the information presented in this book can be usefully applied across a wide range of industries.[12]

Structure of the book

This book is presented in three main parts. First, there is the introduction to project management, its impact on society and the idea of the project life cycle, while the second part deals with the phases of the project life cycle. Part three looks more closely at the activities that must go on all the time in order to move the project sequence along.

The reader will notice some overlap of ideas between chapters. This is inevitable as I seek to reinforce concepts that apply to more than one area of project management and examine issues from different perspectives.

Part One

Overview

1

Chapter One

1

Project Management and its Social Impact

The following definition is not perfect but allows us to make some progress.

A project is a one-off change to be achieved by a finite, time-ordered and interrelated set of tasks. The one-off change is the project; the time-ordered set of tasks is called the project sequence. Project management is the identification of the one-off change and the management of the project sequence.

This definition is different from that adopted by many, and has been chosen so as to define the word 'project' in quite an explicit way. Having done that, project management is then defined in relation to acquiring or achieving the project, going beyond the use of such project management techniques as CPM or WBS. This definition allows you, when you have a project, to practise project management using techniques and methods drawn from both the area of general management and from the area of what is now recognised as project management.

The factor that distinguishes the problems handled by general management from those handled by project management is the one-off nature of projects—one-off means that there is no repetition, once the project is finished, it is not repeated. A one-off event is related to experience—if you have not experienced something before, even though others have, then it is a one-off experience. While a builder sees building a house as a very normal repeating activity, a new home owner will see it as a real one-off experience.

Process versus project

In theory, one can put forward the position that there are two distinctly different types of management situation. In one, activity is repetitive, just keeping the plant turning over producing the same product day after day. In the other, there are one-off management objectives, such as writing a new software program.

For instance, a project is not the management of a chemical production process where the work is mainly directed towards keeping the process going and monitoring progress; however, the installation of new machinery would be a project. Running a rail service where the main objective is to keep the trains on time is not a project, but introducing a new train service is.

Where an activity is repeated, there is the possibility of improving performance the next time round by identifying trends, by improving the process and so on. In quality terms, the activity is amenable to quite well-recognised continuous improvement techniques. However, with the one-off activity, when it's finished, it's either worked or it hasn't. If it hasn't, it is sometimes but not always possible to repair the damage. It really is important to get it right the first time. Examples that could fall in the one-off category are constructing a building (a well-known activity often leading to lots of undesired repair work!), reorganising a company (mistakes here can be fatal) and building a ship (a bad design could lead to a sinking).

The term 'process management' will be used to describe the management of the repetitive activity while the term 'project management' will be used for the management of the one-off activity. General management has oversight of the whole business and is an ongoing management activity which, depending on the situation, may adopt process management or project management or both. Typically, one runs one's life or one's company in the general management framework and draws on process management or project management as appropriate.

Combining the techniques

In practice, there is no nice clear distinction between one-off and repeating problems; real management problems present as a mixture of both. Issues that are generally managed according to process usually have some aspects of project management mixed in, and vice versa.

General management is, in loose terms, those approaches embodied in typical Master of Business Administration (MBA) courses around the world. Many of these courses have an underlying model of stability or of repetition, hence process management is often covered quite well. In the actual delivery of these courses, change is constantly dealt with (in discussions on investment decisions, for example) but often without making the connection with project management. Thus the delivery of these general management courses ignores a whole range of methodologies, the project management methodologies which concentrate on the management of one-off change.

This book takes the view that project management theory is an extension or a development of the long line of thinking that developed into process management theory. The decision to open a new shopping centre is commonly made by people who see themselves as general managers, not project managers. Developing a new shopping centre is, on the one hand, a one-off venture; it is a new development in a new area, appealing to a new set of customers. On the other hand, it is repetitive in that the developer uses experience gained in earlier ventures, on other shopping centres. When it comes to the building of the shopping centre itself, a recognised project activity, one finds that it actually involves lots of repeating or process activities—one shop type is repeated over and over.

Running a train service, as mentioned earlier, is not a project but a process. However, there are many one-off activities such as upgrading signals, introducing new train timetables and introducing new repair and maintenance systems. However, running a train service has special problems to deal with, such as making contingency plans for crashes. A train crash occurs more than once; in fact, if you run a train service you can expect a crash every now and then. Although one has to accept the fact that a crash will occur, one doesn't know when or where. This represents a very interesting form of repeating problem containing elements of both process and project management. On the process management side, policies and procedures can be prepared well in advance and modified in the light of experience; on the project management side, skills in dealing with short timeframes and in managing experts outside one's skill-base will be needed.[13]

Most management situations come with both types of activity embedded in them, and one needs to set up a general management framework to manage both. The author believes that in most cases the problem facing management is of a one-off nature and that every now and then a special case turns up, a repeating problem. Thus a great deal of general management should be project management—this is the opposite to the commonly held view of management.

The project manager needs to be competent in both general and process management, know their techniques and areas of speciality, such as finance, marketing, planning, production, organisational theory, industrial psychology and Pareto methods. These competencies are added to the techniques of project management to form the basis of the discipline of professional project management. Some of these techniques are critical path and other network techniques, work breakdown structure and cost scheduling methods. Project managers should, therefore, have at their disposal the techniques of general and process management, the techniques of project management and their own cognitive processes with which to tackle the challenge of the management of projects.

Identifying the end-point

Many definitions of a project contain the concept of a beginning and an end, but nobody knows when projects start, so the idea that there must be a beginning is somewhat redundant. What is important is that somebody has recognised that the work has an end in sight. It is the author's view that the moment this is recognised is in fact the beginning of the project sequence and signals the need to apply project management thinking.

One should note that where activity is required in a one-off effort to maintain the status quo, we also have a project and a project sequence. Business managers are often confronted with the need to launch projects to maintain an existing condition, such as market share. Efforts to maintain things as they are, which are one-off and finite, are project sequences leading to the achievement of a project.

Managing a sequence

The word 'sequence' rather than 'process' is used to allow discussion of process management as separate from sequence management. The use of the word 'sequence' is to recognise that a new approach to quality is required. The literature on quality discusses the process of production, usually a repeating activity which allows considerable application of statistical tools. Process is a central and important concept in quality management and very important concepts such as process capability turn up in quality management.[14] These concepts are also crucial in project management, but in order to clearly indicate that the area under discussion is project management, the word 'sequence' is used instead of 'process'. Later, we will find the term 'sequence capability'.

Project and project sequence

In the public arena, and among project management professionals, the word 'project' has two meanings. Initially, any distinction between the two meanings is unimportant but later, particularly when the work is expanding in the design and implementation phases, there is a need to be clear about the meaning.

One meaning of the word is the one-off change such as the software package produced as a result of a lot of work, or the restructured organisation achieved after a lot of work, pain and politics. This meaning, then, refers to the end-product or one off-change as the project, and is the meaning adopted here.

In the other meaning, 'project' means the work itself, the work involved in producing the outcome or the one-off change; for example, the actual work involved in producing the software package or the work, pain and politics required to achieve the restructured organisation. This book applies this meaning to the term 'project sequence'. Therefore:

The word 'project' means the one-off change that is to be produced; it is the outcome of work, not the activity. The activity is the project sequence; one could say that the project sequence produces the project.[15]

This definition deliberately separates the objective to be achieved (the project) from the work to be done (the project sequence), a distinction that will be maintained here. The key part of the definition is that a finite, time-ordered set of tasks is required to achieve the one-off change. In effect, the project cannot be said to exist without a preceding project sequence.

Characteristics of project management

It is not necessary, but it would be attractive to be able to start with a definition and work towards a set of logical implications. The most well-known example is probably Euclid's *Geometry*, where he starts with some very basic definitions and develops a whole range of logical implications—properties of triangles, circles and so on. It is not possible, or at most only to a very limited extent, to achieve this in the social and management areas of theory. Some of the features that seem to go with project management are listed below and the reader is invited to think about logical links between these and the definition of project management.[16]

1. Continual elaboration

A key issue for project management is the one-off change. The definition of a project does not require that the nature and form of the one-off change be known at the beginning; in fact, it may not be known until the end of the project sequence, or even later. The definition only requires that a one-off change is to be achieved and that this change arises from the finite, time-ordered set of tasks. Exactly what will constitute the project is uncertain. Only right at the end when everything is done, and in some cases even later, is the full nature of the project seen. These issues are clarified as the project sequence proceeds.

Thus project management implies the management of a problem-solving sequence over time, a continual elaboration of the content of the project and of the project sequence.[17]

2. Brokering agreement

This continual elaboration and problem-solving will involve people who can contribute to the identification of the criteria that will determine the form of the project. Projects always arise from something to do with people—these people may be visible or may be hidden behind organisational structures.

There is no requirement that people agree on the criteria that define the project; usually they don't. In the process of continual elaboration, the project manager will be involved in brokering a solution to which everyone agrees. Brokering an agreed definition of the project is a key project management function.

3. Establishing the criteria for completion

In searching for the definition of the one-off change, the project manager searches for the criteria by which the one-off change can be identified. Identification of completion is a key function of the project manager.

The issue of when the project is complete is very important and must be explicitly managed. At least it will allow people to pay final accounts. The project manager and client need to establish the criteria by which completion can be identified. In essence, continual elaboration must continue until this point is reached.

4. Managing time and resources

Time is included in the above definition. The project sequence is described as a finite, time-ordered and interrelated set of tasks. When it ends may or may not be known for a long time, but it is assumed it will end and it is expected to end (although in many cases it might appear to be never-ending).

It is important for the project manager, immediately after appointment, to come to grips with what is wanted and how to get it. Recognising the time sequence is key. Time presents a framework for the management of the project sequence, and also becomes a driving force. The management of time by networks such as critical path techniques is characteristic of project management.

Management of resources assumes particular significance in project management: it is not only the amount of resources that is key but also their availability and management's ability to use them effectively. With a regular repeating process, one goes on making adjustments and improvements, but with project management, one is first rushing to build up resources and then rushing to reduce them.

As in general management, the availability of resources influences the definition of the objectives to be achieved, and project managers are usually involved in balancing what is hoped to be achieved with what actually can be achieved. But in project management the pressure is heightened and decisions need to be made in a shorter timeframe than in process management.

5. Negotiating trade-offs

Because of the limited or finite resources by which objectives are to be accomplished, there is the implication that trade-offs must occur. This is a major dilemma in managing the definition of a project. While many of these trade-offs are agreed in the planning stages of the project sequence, they will need to be faced throughout it.

Trade-offs go against the ideology of project management. Practitioners, at least in their self-promotion, talk of delivering the project as agreed, in budget and on time. This can lead some project managers to have quite unrealistic expectations of what they are trying to achieve. Trade-offs are to be expected and managed.

Trade-offs imply negotiation, certainly in project management. This negotiation will also be constrained by time and will often take place against a background of great uncertainty, so it would be worthwhile for project managers to develop their negotiating skills.

6. Managing subcontractors

Because projects are one-off events, it is rare for any one organisation to have all the necessary skills in-house for a particular project. If a project can be done completely in-house, the project manager's work is usually made easier. However, here it is assumed that the project need subcontractors.

In selecting subcontractors, skilled organisations or service providers need to be identified. Then the nature of the relationship between the project manager and those providing the service needs to be established. There is also the need to supervise, and take corrective action where necessary, the delivery of the product and service. What complicates these seemingly straightforward activities is that the work of the subcontractors is often beyond the skill-base of the project manager.

7. Coping with unfamiliar technology

Complete familiarity with all the techniques and technologies associated with the project and the project sequence activities is impossible for any one person, so one needs to take advice. In fact, selecting the correct source of advice is crucial. This becomes much more difficult if the sources of advice are very limited in number or only available through the subcontractor.

Dealing with issues beyond one's comprehension is quite unsettling. A dilemma facing many project managers is whether or not to reveal their ignorance. Pretence is usually seen through—the way is to ask questions. Skilful questioning from someone who does not claim to understand can be quite productive and lead to substantial credibility.

8. Cutting across organisational lines

The project manager rarely has formal employer status in relation to all the people working on the project, particularly the employees of the subcontractors. The lack of this formal link does not remove the need for the project manager to manage others who are outside his or her direct authority. This is where there is a significant need for that elusive quality called leadership or people management skills to maintain people's commitment and willingness to support the project.

9. Managing complexity

Although this area relates to the comprehension issue, it goes beyond it. Usually the complexity of the task means that many people have to interact in order to decide what to do and then to implement that decision. The faint-hearted should take courage from the earlier observation that there is both continuous elaboration and problem-solving *over time*.

Skills needed to manage a project sequence

A very wide range of skills is needed, not all of which can be found in a single project manager. The saving grace is that while these skills may not all be wrapped up in one person, they can be found among other members of the project team. There is an overlap in the skills of process and project managers. Without distinguishing between these skills, in managing a project sequence, a project manager needs the ability to:

1. Identify outcomes and deliverables

There is an overriding need to determine what is wanted in a project. The project manager must evaluate the needs and wants of various parties and then be able to maintain clarity on what will be delivered. Some expressions describing this skill are 'delivering what is expected', 'able to determine wants and needs', 'managing the scope of the project' and 'managing change'.

2. Devise a sequence to achieve the goal

Knowing what is wanted leads directly to deciding how to get it. Usually there are a number of ways to achieve a goal and these need to be identified and evaluated. Some expressions describing this skill are 'able to identify necessary strategies', 'identify required resources' and 'identify required sequencing'.[18]

3. Assess the capability of the sequence

This is a central quality question. Having decided what is wanted and having chosen a path or sequence of activities to achieve it, how confident can one be that the desired results will be delivered? It should be kept in mind that the chances of a successful outcome depend on both the sequence chosen and its ongoing management. Some expressions describing items to be considered are 'sequence capability', 'reliability', 'risk management' and 'subcontractor capability'.

4. Recognise if and when the outcomes have been achieved

There is the need to know if the required work has been done. One will also need to know that work has been done to the required standard. On the surface, this seems easy but it is actually a very difficult area in project management. Some terms dealing with this area are 'commissioning and handover', 'approvals testing' and 'retentions'.

5. Deal skilfully with people

Project sequences require considerable interaction with those involved. Clear lines of communication along with morale must be maintained. Much of the work must be done in the context of teams and a wide variety of people will be involved, from very senior executives to close to the lowest level of operator. Terms include 'listening', 'team-building', 'managing industrial relations', 'maintaining political support' and 'obtaining commitment'.

6. Manage unfamiliar complex systems and technical specialists

Project sequences are usually multifaceted and require a wide range of skills. This requires the project manager to manage specialists from a wide range of disciplines, some of which will be other than that of the project manager. Terminology includes 'the consultants advise ..', 'second opinion', 'solve problems over time', 'manage skilled designers' and 'larger than anyone can comprehend'.

7. Identify key issues within large volumes of data

Project sequences generate a huge volume of data, most of which is unimportant. However, some of it is crucial and central to the good management of the project. Unfortunately, the important and unimportant data are mixed together. At any one time on a project, there will be tens and possibly hundreds of issues demanding attention. In reality, the project manager can only give proper attention to about five issues at any one time. Paying attention to just a few issues, however, risks missing a key one, so, to keep on top of the situation, the project manager needs to expend effort in finding the key issues. The risk of missing something is a dilemma for the project manager, but in practical terms is one that must be taken. Important terms are 'Pareto', '80/20 rule', 'can't see the wood from the trees' and 'lift one's gaze'.

8. Respond to contingent situations

The work of creating a project is rarely a steady-state activity and the project manager is likely to face changing and unexpected situations. Project sequence activity tends to jump about. The general experience of project managers is one of constantly trying to keep the project sequence on track, and often trying to pull it back on track. Plans will always need to be changed or adjusted, something always crops up and so on. Project work requires flexibility to be able to respond to the here and now, a particular mental framework and the ability to quickly work out a solution to a particular problem.

9. Identify and manage interfaces

The management of the interfaces is clearly the responsibility of the project manager. Maintaining an overview and managing the interrelationships between the parts of the project is managing the interfaces. Interfaces are ubiquitous. They occur where the lines of responsibility between the consultants meet (or do not meet!), where the lines of responsibilities between the subcontractors meet (or don't), and where the lines of responsibility of the team members meet (or don't). Common terms include 'the right hand doesn't know what the left hand is doing', 'clashes of services' and 'falling between two stools'.

10. Manage time and cost

Project managers need the key project management skills of managing time and cost in one-off situations. Terms used here include 'bringing the project in on time and on budget', 'contingency', 'overruns' and 'time and cost budgets'.

11. Set up and run the project management information system

The project manager will have to set up a new office, essentially from scratch. Decisions will need to be made on how data is to be managed and how it is to be captured, analysed and stored. This will all have to be done anew for each project. Previous experience in setting up information systems would obviously help. Terminology includes 'document control', 'databases', 'filing' and 'record management'.

12. Manage political and community issues

Project management literature is now identifying external political and community issues as strongly influencing outcomes—in many cases they are seen as a major threat. Projects bring change in their wake and people's lives and interests are affected, so strong forces exist. Terms include 'public relations', 'impact statements' and 'managing stakeholders'.[19]

13. Manage contractual matters and contract strategy

Projects usually involve many contracts; if it is international, then the contractual matters are more complex. On the basis of advice, the project manager will have decide how contractual matters are to be handled and how the ongoing contract is to be dealt with. Commonly used terms include 'managing subcontractors', 'agency versus principal', 'liquidated damages' and 'rise and fall'.

14. Maintain a sense of urgency
The project manager will need to drive the project, to keep it going, to push the project sequence to the next step. This requires energy and enthusiasm for the progressing of the project sequence.

15. Cope with risk
Projects are risky, some more risky than others. Project managers need to come to terms with the fact that their profession always has a level of uncertainty associated with it. It requires psychic energy to keep dealing with uncertainty. However, rather than adopting a head-in-the-sand approach, one should actively set out to identify the areas of danger.

Social impact of project management

Project management starts off in organisations as a simple idea—put somebody in charge and let them focus on the project, free of distractions from other organisational activities. But this simple idea, when it is used more and more often, leads to quite profound social changes. These changes will be manifested in changes in organisational structures and changes in career paths. They will also lead to changes in what is valued, and what is not valued.

To summarise what is presented below, project management leads to a single point of control; satisfying, rather than optimising, organisational forms that tend towards temporary flatter hierarchies and groups; a concentration on co-ordination rather than functional skill, more autonomy and, in many cases, a two-boss situation. It is a culture change, not always a pleasant one, but one that is very seductive to project managers.

Changing technologies
Underpinning rapid change is technological change. The development of a new product is the first evidence to the general public that there has been a technological change. A modern example of this might be the microchip (an older example is the invention of clocks), which allows a computer to be produced. So the new product emerges.

This product then spreads to other industries (and countries), before becoming incorporated into the production process itself. As we all know, the computer is now part of the production process of a whole range of products. Once the technology enters the production process, there is labour displacement. The technological change not only alters modes of production, it alters society. The demand for some skills decreases while the demand for others increases. Whether or not there is an increase in overall demand for labour is a moot point;

however, the people displaced are not the same people who experience an increase in demand for their skills. When the electronic watch was introduced, it led to an increase in watch-making in areas other than Switzerland.

Technological changes interact and combine to provide the opportunity for even further and more profound change. The existence of the motor car, coupled with the existence of the fridge for food storage, for example, is a major factor in the rise of shopping malls and the decline of the neighbourhood shopping area, because one can move to less frequent shopping and do it further afield. Putting telecommunications and computing together provides an enormous platform for change, from electronic data interchange to Eftpos, to mobile phones and the Internet. The combination of changes does not appear to lead to a slowing-down in the process of change but the reverse. The more rapid rate of change leads to the need for changes in management.[20]

Meeting the challenge of change

The nub of the argument here is that change puts pressure on the co-ordination of the work of specialists. Specialists have always had to be co-ordinated and in stable, repeating processes, specialists were co-ordinated using the functional hierarchical organisational structure.

Most people are familiar with the organisational chart that has the board of directors at the top, above the chief executive or general manager. Below that is an inverted, tree-like structure going down to the operator at the lowest level. Basically, it shows who's boss of whom; this is a hierarchical structure.

Besides having a hierarchy, skilled people of a particular discipline are collected together in departments or subgroups, for example, the engineers and the programmers. This is the functional breakup of the organisation. When we combine the hierarchy with the departmental breakup, we have a functional hierarchical structure.

When a change is in progress, the functional hierarchical structure is very slow and inefficient in co-ordinating the specialists. When changes start to come at a faster and faster rate, the functional hierarchy breaks down and becomes ineffective, so a new form of organisational structure is required. In most cases, that structure has more flexible relationships between people with less emphasis on the hierarchy. Single points of control emerge at lower levels in the organisation. Typically, those single points of control are project managers exercising the function of project management.

Managing in a functional hierarchical structure

Functional hierarchies have all sorts of rules, such as who may make a decision, who may direct who, and many formal ways of communicating. Besides the formal ways there are also informal ways of doing things. There is a formal structure and an informal structure in all organisations.

The organisation chart combined with the written rules (and also possibly with some of the unwritten rules) is called the formal structure, which will have the procedures for decision-making. In the functional hierarchical formal structure, information can only be transferred between departments through their respective heads. Thus co-ordination between departments requires that information and data moves up and down the organisation.

Also in existence, but not as visible, is the informal organisation. This is the whole range of personal contacts and relationships that exists between people in the organisation but which is outside the formal organisation. It includes the old school tie links, the grapevine, blood relationships, friendship and hate networks. This informal structure actually does considerable co-ordination work. For example, the marketing and salespeople may contact their friends in engineering to sort out some problems or gain some favours for their customers.[21]

When problems fail to be resolved in the formal structure, as they inevitably will during periods of rapid change, they are given to the informal structure which may or may not succeed. An example of managing a problem in the informal structure might be a manager going outside his or her authority to approve an expenditure, hoping that the necessary paperwork will be sorted out later. But handling problems in the informal structure is not a proper longer-term solution and indicates the need to change the formal structure. Managing projects in a functional hierarchy by means of the informal organisation signals the need to move to a structure more in line with project management. It should be said here that the project management organisation will also have its formal and informal structure, and it too can break down.

While recognising that functional hierarchies are more complex organisational structures than the model just described, this model is at least sufficient to get a perspective on the problem. Functional hierarchical organisations represent insulated and separated groupings of expertise—they maintain expertise, but each department tends to focus inwards on its own expertise without focusing on the co-ordination of effort needed to achieve an objective outside their own area. This type of structure is efficient and functional where the work is constant in nature and essentially may be regarded as repetitious. While it is appropriate for the operations side of a railway to be a functional hierarchy, it will not help in the introduction of the very fast train. It is appropriate for the chemical production side of a chemical company to be a functional hierarchy but not its research department. When the job has an end-point (and particularly when the exact nature of that end-point is unclear), the functional hierarchy breaks down. At best, it is inefficient; at worst, it is a major barrier to success.

Managing in a project management structure

The organisational forms that are more in keeping with project management have certain features, but before discussing those features it is worthwhile pointing out a fundamental dilemma in choosing an appropriate organisational form. This dilemma relates to the need to develop specialist skills. While rapid change leads to the breakdown of the functional hierarchical structure, there is still the need to develop specialist skills. Specialisation is developed and protected in the functional hierarchical structure, and not in the project management form of organisation. Thus the dilemma—co-ordination requires the breakdown of the functional hierarchy while the development of expertise requires its continuation.

Developing specialists skill, spending the money on training, learning from mistakes, developing experience and so on are compatible with the functional hierarchy and with reporting to a departmental manager. These skill-enhancement activities are not compatible with one-off events; the costs need to be distributed across many projects. But getting work done often requires project work, and hence reporting to the project manager. Some interesting attempts have been made to resolve this dilemma, eg, the matrix structure.[22]

Organisational forms that are established for the practice of project management have some or all of the features described below.

1. Single point of control

The idea of control brings with it a host of related words and ideas, such as 'control', 'authority', 'responsibility' and 'accountability'. Some of the distinctions between the meanings are quite subtle; suffice to say that the single point of control means that the power of decision-making is concentrated in one person or entity. In this case, it is the project manager who may delegate part of the control to the project team members. The value in having a single point of control is that confusion is reduced because decisions are better co-ordinated. This becomes crucial when dealing with a situation that is always changing, as is the case in managing the project sequence.

In repeating activities, a practice develops and blends in with the activity at hand. Everyone knows who to go to for particular information, because these people have developed their role to include the provision of this information. However, in the execution of projects, by the time a practice has become standard, it is time to move on. Projects rarely allow the situation to arise where information distribution stabilises.

As the single point of control, project managers are better able to keep track of the exact status of the project, are better able to monitor and control changes, and are better able to control expenditure and generally keep things shipshape. Without this single point of control, the danger of disintegration looms large.

Many functional hierarchies, when confronted by a project, set up a single point of control outside the hierarchy to manage the project. This single point of control then ranges free over the organisation as the project requires. For most projects, the single point of control needs to be outside all of the departments that constitute the organisation. (There is an exception to this; if the bulk of the work of the project sequence has to be done within one department, the work might be co-ordinated by a single point of control within that department.)

Project management reinforces the concept of a single point of control but not absolute power. As we shall see below when discussing the authority of the project manager, the power of the single point of control is limited.

2. Satisfaction of criteria

The functional specialist culture is one of completeness and accuracy. It is one of paying attention to detail and one of doing things more and more efficiently. It devotes considerable energy to doing its part of the job well, using its previous experience as a benchmark and attempting to improve. It is the opportunity to do something again that gives rise to the possibility of optimising what is done.

But the project sequence does not provide much opportunity for specialised learning. In project management, the objective is to get things done within a set of constraints, some constraints being more important than others. Whether or not these represent the optimal set of constraints is not the issue facing the project manager—what must be satisfied are the constraints set down or agreed. What this means in practice is that a project manager will be under considerable pressure to bring a project in under budget, but not necessarily for the lowest possible price. The objective is to satisfy a constraint, not produce the optimal result.

There is often a simple logic behind the concept of satisfying rather than optimising. An example might be a project to construct a shopping centre that must be open for the Christmas shopping period. This is the most important sales period, dominating the minds of shopkeepers. If one misses the Christmas period, one might as well miss the rest of the year. The project manager will feel under enormous pressure to meet the deadline.

The project management mind-set is different from that of the functional hierarchy. In a functional hierarchy, one can attempt to improve a job, one can spend time doing it as well as possible. In projects, one has to do the job as well as possible within severe time constraints, and this can lead to one not doing it as well as one might have liked. Some will find this distressing, particularly if they are competent enough to see how much better a particular activity or job might have been, given more time. The shopping centre could well have been more beautiful but if that meant it would be late for Christmas, less than perfect will have to be accepted.

In practice, what happens is that one or two criteria are identified as critical. One will then try to satisfy them as much as possible within the constraints of the other criteria—'We will bring the project in as soon as possible within cost and other constraints'.

Later, we will come across this issue of satisfying rather than optimising in other guises, such as the guise of efficiency versus effectiveness—just getting the project done is more important than getting it done efficiently. It will also come up in the issue of timeliness versus accuracy in relation to management data. One will have to accept situations of suboptimal decision-making. To have to satisfy rather than optimise permeates project management and its mind-set will be anathema to many specialists.

3. Flatter hierarchies and groups

There is an inherent logic pushing project organisations away from the deeper hierarchical model to the flatter structure. That logic derives from the need to process information on projects in an interactive rather than in a sequential mode.[23]

Processing data in a sequential mode means taking an amount of data and working on it; more data is then taken in which does not significantly alter one's view of data collected previously or significantly alter the nature of the next lot of data to be collected. The important thing here is that with one amount of data one can do a considerable amount of work before needing the next package of data. The question 'Which is the next train to be fuelled?' leads to the identification of a train; the work of fuelling can then go ahead without knowing the status of the following train. This kind of non-interactive sequential processing of data is possible when running systems where there are considerable levels of control over what is being produced, where people already know much about what is being done and about what is expected to be done, where targets can be more easily set and reached, where the factors that might limit production are more easily identified and coped with, and where there is considerable certainty about what is being done. Many of these features are present in process production.

Processing data in an interactive mode means that someone passes a piece of data to someone who, on the basis of the data received, responds with further data which in turn gets more data in response. The data requested and exchanged is influenced by what has been delivered previously. A simple example here is a visit to the medical doctor; the patient tells the doctor how he or she feels, to which the doctor typically responds with a question, and more data on symptoms is given and so on. What is happening here is that the content of the data passed across depends on the data already given. In other words, the process of solving the problem requires one or more people to interact with each other. Typically, this form of data processing is managed in a face-to-face mode; it is incredibly slow otherwise. Unit production typically requires processing in the interactive mode and hence calls for much face-to-face interaction. Projects fall into the area of unit production.

Projects require considerable processing of data in the interactive mode. All the key issues and much of the detailed problem-solving must be discussed face-to-face—engineers must speak directly to architects, software designers must talk to users, film directors and actors must consult and so on.

The requirement for face-to-face interaction leads directly to shallow hierarchies and groups. The less that is known about a project, the greater will be the requirement for face-to-face interaction and the flatter the hierarchy will be. What will be seen is that as the project work proceeds, as more and more data is gathered and organised, as more and more is known about the project, the hierarchy gets deeper and deeper. Hierarchies at these later stages are only sustainable through development of knowledge of the project. When designing a scene for a film, for example, only two or three people may be involved, but when the scene details are known and the method for informing people is worked out, then, and only then, can many more people become involved and be organised in a fairly deep hierarchy.

4. Co-ordination takes priority

It is essential to understand that certain responsibilities require explicit attention. These responsibilities are to:

- ensure good communication of plans
- clarify and gain agreement to objectives
- ensure adequate work is done in developing plans
- maintain control of staff
- ensure all work is properly authorised
- resolve conflicts of priority, and
- maintain morale of team members.[24]

These are key responsibilities of the project manager and cannot be left to wait—the project sequence activity does not wait. In a process management situation, many items can be left to be picked up later, but in project management, the timeframe is compressed. All of them are co-ordination activities. From the project manager's point of view, the issue of co-ordination is central; the specialist skill is put aside.

The extent of the project manager's responsibility for technical failure is debatable.[25] Certainly, it is not expected that technical problems should be solved by the project manager, but the project manager might have been expected to set up a system and co-ordinate a process to deal with them.

There is ongoing debate about the extent to which a project manager can manage without a basic knowledge and understanding of the underlying technology, that is, without an understanding of a specialist skill. This debate comes up particularly in discussions about the nature of project management. The author believes that project managers should be able to quickly read themselves into the particulars of the technology: it is essential that they come to grips with the technology, but it is not clear to what level.

There are two issues here. The first is how much a project manager should know about a technology; the second is the process whereby a project manager can quickly and effectively get a grip on it. The first point is unanswerable, but not knowing anything should not deter project managers from venturing in as long as they recognise that they will need to learn. For their psychological survival, it will be useful for them to remind themselves that the data can be picked up over time; it does not need to be learned immediately. As time goes on, project managers pick up more and more of the vocabulary, the insider language and the concepts.

But when all is said and done, it will be the responsibility of the project manager to co-ordinate the work of the specialists, not to do the specialist work. The debate about to what extent the project manager needs to know the technology boils down to the amount of technical or specialist skill knowledge that is required in order to be able to do the co-ordination.

The emergence of the importance of co-ordination points to a more profound change going on in society—the increasing reliability of technical systems. The author believes that, as the practice of a specialist skill becomes more reliable and more amenable to being governed by rules and codes of practice, its value declines. It declines in both economic terms and organisational standing. Looking at it from another angle, as technology becomes more reliable, its status declines; technology is no longer the key but its co-ordination becomes central. If the technical systems themselves were not reliable, the co-ordination function would be a low priority and almost irrelevant. The recognition of the 'specialist in co-ordination' is based entirely on reliable technology. Once a technology becomes reliable, both it and its practitioners seem to lose their importance, their status and their power to command attention.

On the negative side of project management is the fact that once a technology becomes reliable, its practitioners become more open to negligence law suits. The management of technology then becomes the co-ordination of a defensible framework of practice. Many projects are weighed down by paperwork whose only purpose is to assist in the defence in a possible lawsuit, not expedite the project. This is emerging as a considerable burden; partnering is seen as one attempt to alleviate it.

Project management usually involves appointing subcontractors; projects that do not need subcontractors can simplify matters. Subcontractors may be professionals providing specialist advice or companies who provide particular services and products. The project manager's function is to co-ordinate their inputs as well as make certain judgements about their performance.

Project co-ordination turns up in many guises; some of its functions exist under such titles as liaison officer, expediter, special teams and project co-ordinator. The function of co-ordination is seen by some as a shallow activity; however, as projects get bigger and more technically complex, the co-ordination activity becomes an activity demanding a wide range of high-level personal and technical skills.

5. Authority and responsibility

Many believe that management and control go together. This is an illusion. Quite often, probably in the majority of situations, people are managing without having control. Many spouses manage the family money without technically having control over it. Project managers routinely face the difficulty of not having the level of authority required to match their responsibilities. While they may be able to identify the decision to be made and be capable of making that decision, they very often do not have the full formal authority to implement it. Project managers call for this to change—it probably won't. They are unlikely to gain the level of authority they need and there is little use in complaining about it; one has to find defensible ways of operating in such an environment.[26]

It will not be changed because the levels of control and decision-making required in project management usually vastly exceed the authority of the most senior executives of the organisation. Projects often involve the expenditure of millions of dollars and involve making commitments of such magnitude. In a $20 million project there will almost certainly be a few subcontracts worth at least $1 million. Few senior people have this level of delegated authority and there will therefore be considerable resistance to giving such power to project managers.

But the issue of authority goes beyond money matters. It extends to dealing with the people working on a project who are typically only on temporary assignment. They are usually formally employed by organisations not associated with the project. To follow formal procedures, the project manager would be required to go through the hierarchy of the employing organisation in order to issue an instruction or a request. In practice, the project manager deals directly with the employees. Although the project manager has little formal authority over them, their co-operation in achieving the project must be obtained.

By way of example, the design phase of a building involves architects and engineers. The formal reporting procedure might involve the engineering designer reporting to the managing partner of the consulting engineering company; the architectural designer reports to the managing partner of the consulting architecture company; and the project manager reports to the managing director of the project management consulting company employed by the client to manage the design phase.

In the formal contractual arrangements, there might be a legal relationship between the client and the project management consulting company, the client and the consulting architecture company, and the client and the consulting engineering company. There is no formally documented relationship or agreement between the project management consulting company and the consulting architect company, and none between the project management consulting company and the consulting engineering company, and none again between the consulting architect company and the consulting engineering company.

Whole relationships are moved one step further when the project manager, in person, must work with the people doing the work of the architectural designer and the engineering designer. There are no formal agreements here, there are no formal authorities here at the personal level—they are relationships mediated through a range of contracts and understandings.

The formal authority process would probably take the form of the person who is the project manager passing an instruction to the managing director of the project management consulting company, who then passes it on to the client, who passes it on to the managing partner of the consulting architects, who passes it on to the architectural designer. The architectural designer then requires more information and passes it back to the managing partner of the consulting architects, who passes it back to the client, who passes it...and so on. This is clearly not viable.

There is a need for the kind of direct contact outlined in the interactive problem-solving situation. The point here is that while efficiency demands an interactive problem-solving approach, on most projects the authority to execute does not formally exist. It happens in practice on the basis of some common understandings, not on some strict, formal, strong legal basis. Project managers must gain and hold authority by their maintenance of the common understandings and by the qualities of their interpersonal behaviour. The authority is on an interpersonal foundation rather than something more substantial. The interpersonal foundation is probably the most efficient, but it has its risks. This is why some project management practice looks like an exercise is managing things so that someone else is always to blame.

There is another quirk in this issue of authority. The person occupying the position of project manager will have a formal position within an organisation. This formal position will be seen as having an associated status. But often the project manager will have to gain the commitment and support of other people of higher status in the organisation in order to carry out the project. Lower-status people in one organisation will have to direct higher-status people in other organisations (the architectural designer may have to give directions to the managing partner of the consulting engineering company). The project manager no longer remains within the bounds of the hierarchy but is expected to deal with anybody who has an impact on the project. The good of the project becomes the justification for moving outside the functional hierarchy, it makes actions legitimate and is what overturns the niceties of the formal hierarchy. Managing this role requires considerable tact and finesse.

What the project manager lacks in authority he or she must make up for with interpersonal skill.

6. The two-boss situation

The two bosses are the project manager on the one hand and the head of the department in the functional hierarchical structure on the other; for example, the systems analyst must report to both the project manager and the manager of the IT department. This two-boss situation arises quite naturally in the project environment and there are many ways of dealing with it, the matrix organisation being one.

The two-boss situation arises fundamentally because there is a direct tension between the need to focus on the project and the need to maintain one's specialist skills. It does not apply to the project manager; it applies to the specialists working on the project. However, the two-boss situation is something the project manager has to contend with.

Projects are temporary events. People are usually employed in fixed positions (this is changing) and usually in functional hierarchies. The specialists are employed in the functional departments—the engineers in engineering, the salespeople in sales and so on. When a project comes along, there is a call for resources. If an engineer is allocated to the project, who is the boss—the project manager or the head of the engineering department? In fact, they both are but for different aspects of the employment.

If the project demands the full-time effort of a salesperson, then that person will join the project team and, for all intents and purposes, the boss for the duration of the project is the project manager. The situation becomes more complex if there is not enough work on the project for full-time effort. In this case, the salesperson's working time is divided between a number of projects and maybe also on some departmental business. Whoever is head of the sales department is boss, but there's a whole raft of project managers who also act as bosses.

It is vital, for the sanity of all concerned, to be quite clear about who is the boss of what. The head of the department is responsible for the professional skills delivered by the project participant, while the project manager is responsible for the time usage of those skills. In the case of a number of project managers, each is responsible only for the time usage in their specific area. This division of responsibility is not without difficulty.

The two-boss dilemma is here to stay, no matter how much management writers call for the adoption of the 'one job, one boss' principle. It will lead to more uncertainty and calls for more flexibility from both the supervisors and the supervised.

Responding to culture change

The introduction of these changes to an organisation often causes severe culture shock to existing employees. The changes mean:

- a move from general control to focused control because of the emergence of a single point of control;
- quite a different view of career paths because of the move from deep hierarchies to shallow hierarchies or to matrix structures;
- a move to value urgency because of the focus on deadlines, a move to what can be described as a 'time fetish';
- a move to value immediate relevance because later data may be too late to be used;
- a move to developing satisfying rather than optimal solutions because of time and other constraints;
- concentration on co-ordination rather than functional skill because of the reliability of technology;
- more autonomy (possibly);
- in many cases, a two-boss situation; and
- more outsourcing, because of the rise of subcontracting.

Chapter 2

2

The Overall Management of Project Acquisition

Project management is about acquiring, getting or achieving a project. The words 'acquiring' and 'getting' easily apply to physical projects. Where the project is an organisational change or some similar non-material outcome, although the word 'acquire' is perhaps not so appropriate, we will continue to use it. In this context, the word 'acquire' means 'bring about' or 'cause to happen'.

One of the main management strategies is to achieve the project by a set of phases rather than by a one-off, undifferentiated sequence of activity, which is more suited to very small projects. The vast majority of projects require that the project sequence be broken down into stages or phases known as the 'project life cycle' strategy. This strategy is widespread, almost universal, in the practice of project management. Choosing and managing the project life cycle is the major component of the work of the project manager. In this chapter, the overall approach to the project life cycle strategy will be discussed; a detailed examination of its components follows in later chapters.

The work being carried out in the project life cycle needs to be controlled. The usual control method, the cybernetic model of taking measurements and then making adjustments, will be described in some detail in this chapter. Combining the cybernetic control model with the life cycle model allows a system of control to be set up whereby the objectives of the subsequent phases are set up in the previous phase. This is a powerful method of ensuring that what is being delivered is in fact what is expected.

Project life cycles

It is important at the outset to see the project life cycle as a management tool and not as something intrinsic to projects. Project sequences may or may not have life cycles. Saying that a project sequence has a life cycle suggests that project sequences naturally go through phases—this is inaccurate.

The project life cycle is imposed on a project sequence by management so as to make it easier to manage the project sequence; it is an artificial device used by management to gain control of the sequence of achieving the project.

The choice of phases will be influenced by the industry in which people are working. The introduction of a new drug to the market has quite a different set of phases from those associated with the construction of a building; there will be common features but there are quite important differences. So in choosing the phases the project manager needs to take account of industry practice and experience.

What is not often recognised is that the procurement of any one project has a number of life cycles, not just one. Project life cycles vary to suit the needs of the project participants. All the different, significant participants require their own project life cycle to be identified and managed; the financier's life cycle will be quite different to that of the marketing specialist's. Thus, more than one life cycle is involved in the management of a project; these life cycles will need to be compatible with one another or managed to satisfy the requirements of each. Different project managers choose different project life cycles. Once identified, the project life cycle is then used to assist project management in keeping control and in making decisions.

Defining and delineating

As a reminder, the definitions given earlier:

A project is a one-off change achieved by a finite, time-ordered set of tasks. The time-ordered set of tasks is called the project sequence. Project management is the identification of the change and the management of the project sequence.

The tasks can be further placed into subgroups of tasks called phases; each subgroup of tasks is given a name, such as the feasibility phase, the implementation phase or the handover phase.

The life cycle is a time-ordered set of sub groups of tasks of the project sequence. A subgroup is often called a phase.

The word 'phase' usually has a time connotation—the words 'during the design phase' clearly indicate a defined period of time. It is probably because of this strong time connotation that the word 'phase' is used in project management; time is almost always a central issue on projects.

In the construction and engineering industry, people often refer to four subgroups of tasks, that is, four project phases: *initiation, design, implementation* and *handover*. In initiation, the project is identified and its feasibility tested; during design, the drawings and specifications for the work are prepared; in implementation, the building is physically constructed; and in the handover phase, the building is handed over to the users.

In new product development, the project phases might be *idea generation, product definition, product engineering, pilot start-up, manufacture* and *marketing*. In the idea generation phase, some very general ideas about a product are developed (eg, the use of baking soda in toothpaste); in product definition, the design of the product is decided (exactly how much baking soda in each packet, how the packet is to look, etc); in the product engineering phase, how the product will be manufactured is decided (which plant and machinery is required, what skilled staff are required, how the mixing is to be controlled, where is it to be made, etc). Pilot start-up is a trial production run to make sure everything is in order and to expose any problems (some would regard this phase as the end of the project, and manufacture and marketing as belonging to another sphere of management, not project management). The product is then manufactured and finally marketed to the consumers.

The precise delineation of one phase from another is a matter of judgement, as is the selection of the number and kinds of phases. It is important not to become attached to any one set of phases or delineation of phases, but to be detached and objective. The phases selected should be those appropriate to the management of the particular project sequence. It should also be remembered that different stakeholders have different management issues to consider, and therefore different life cycles.

Developing the project life cycle

Project life cycles vary from project to project. However, there appears to be a small number of groupings that can be a starting point in the development of the project life cycle for one's particular project. The main groupings appear to be engineering, new product development, software development and organisational change.

Engineering projects are often called 'hard' projects, the word 'hard' implying that there is a fairly definite, mechanistic way of going about them. Organisational change is in the extreme of the 'soft' project category: much is uncertain and how project management applies is not always clear.

One can relate many life cycles within one industry, but relating them across industries is more difficult. When confronted with organising a project in a specific industry, it is probably well worthwhile identifying the usual phases of projects in that industry as a starting point. However, the important point remains—phases are a managerial device, used to help manage the acquisition of the project.

Engineering project life cycles

These life cycles seem to have four main phases. The first is always concerned with what one is going to do, the second involves planning its form, the third is its acquisition, and the fourth is its transfer to the owner. What is interesting about this form of the life cycle is that it is clearly focused on how the owner sees things rather than how the constructor or supplier sees things.

From the supplier's point of view, there must be a phase between when it is decided what is wanted and when the actual manufacture or assembly of the item begins. A building contractor must do some preliminary work after winning the contract and before turning up on site; a turbine supplier must make arrangements before the actual sequence of manufacture starts. Typically, this phase includes finalising arrangements with subcontractors and suppliers. For this reason, this book covers the 'pre-implementation' phase.

The four-phase model is frequently presented ignoring the pre-implementation phase. It should also be noted that these projects are considered to start much later than the author considers projects to have started. This book identifies a 'transition phase' which is not fully considered in the four-phase model.

New product development

These life cycles typically have more than four phases. New product development life cycles resemble engineering life cycles mainly because there is a fair amount of engineering actually involved. However, the practice would involve a wider group of specialist skills than would most pure engineering projects. A big difference from the engineering project is that there are essentially two designs to be done in the new product development project—there is the design of the product itself and there is the design of the manufacturing process. The design of the first probably requires a sensitivity to fashion and elements associated with aesthetics, while the design of second can probably focus much more on functional requirements.

Another difference is that there is probably some opportunity for trial and error. One can trial a manufacturing process and fine-tune it, whereas with an engineering project such an opportunity is remote.

Software development
Often the challenge in software development is to identify what is required, not how to get it. One of the keys to successful software development is to delay coding until the software product is defined.[27] There is a wide variety of arrangements of phases possible here: one life cycle might consist of a customer-orientated phase (what do they want and need), followed by a feasibility phase, then specification development, detailed design, coding and testing, and finally handover. Alternatively, there could be an initial investigation, a feasibility phase, a user-requirements definition, then system design, software definition, coding, installation, and finally handover.

When one is developing a hardware and a software system together, one has to manage it as two parallel projects that must come together at critical points.

Organisational change
This is rather more difficult to lay out as a clear set of phases. The literature often describes three steps, such as 'unfreeze', 'change' and 'freeze'.[28] In this model, one does things that make the organisation more receptive to change, implements changes and then cements in those changes.

Just how the organisational change is to be managed depends on the power of those pushing for change. Groups holding significant power in the organisation can impose change and implement it by decree. However, even in these situations, considerable work is usually done to facilitate the change and reduce resistance. This work can be broken down into tasks, and timed plans set in place. The phases in this case might consist of a data-gathering phase, a planning phase, an implementation phase (which can consist of appointing people), and a post-implementation phase directed towards confirming people in their new roles.

Where those promoting change do not have much power, the project is essentially a political game. While it is still a project, other texts, such as those on politics, might provide better advice than a text on project management.

In practice, most projects, not primarily those directed towards organisational change, have an organisational change component. The introduction of new machinery or of new capital facilities both involve organisational change.

Overlapping of phases and parts of phases
Projects tasks are carried out both sequentially and in parallel. If all the work is carried out sequentially, drawing the boundary between phases is quite simple—one simply looks for appropriate partial completion points and groups the tasks between these points into a phase.

Phases can be set up in parallel. However, the execution of work in parallel leads to quite complex boundaries between phases and needs very careful management. A major parallel-phase strategy in project management is called 'fast-track'.

Fast-track means starting the work on one phase before work on the previous phase is finished. On a building job, one might start to cast the foundations before the details of the structural design of the building are complete.

There are very good reasons why a phase should not start until the preceding one has finished—it reduces one set of risks. However, because of another set of risks and effectiveness requirements, subsequent phases are often started early. Thus it will happen that work from much earlier parts of the project sequence will be going on in parallel with later work. Fast-track leads to quite difficult control issues and imposes further management demands on the project manager.

There is considerable debate in the project management literature and in the practice of project management about the appropriateness of fast-track. Strong views are held on both sides of the argument. However, one thing is clear—fast-track is not seen as a cheaper option (and it may not even save time!).

Project life cycles as management tools

The project life cycle assists in the management of the sequence of tasks needed to complete the project; it helps in the identification of, and decision-making on, issues, and it helps identify the work to be done and when it is to be done. While many are tempted to look on the project life cycle as simply a descriptive device, it is an important decision and control mechanism for suppliers and the various stakeholders in the project, as well as for the project manager.

1. Maintaining an overview

It is essential to maintain an overview of the whole project: the sheer volume of day-to-day detail tends to obscure the overall situation and thus hide potential threats or opportunities from the project management. The life cycle approach acts to reduce the difficulties associated with 'lifting one's gaze'.

To see the overall project and how it all comes together involves defining the overall reach, influence and responsibility of those involved in the project, and recognising the complete amount of work needed to achieve success. For example, in the development of a new product described earlier, is manufacturing part of the project or is getting manufacture started part of the project with a handover sometime after the production has started? Are the industrial relations negotiations to cover the introduction of the new plant part of the project?

As these questions show, the boundaries of a project can be quite wide. Failure in any one of these aspects would reflect badly on the project management. To avoid getting lost in the details and to ensure identification of the boundaries of the project, a project life cycle is essential (another tool is the Work Breakdown Structure).

2. Identifying the tasks

In larger projects, the phases help to identify and elaborate on the project sequence, the time-ordered set of tasks. Usually, the phases are recognised or identified before the tasks are decided on. The definitions given earlier in this chapter would imply that the tasks are identified before the phases; in practice, the phases will be identified as a helpful mechanism in finding and identifying the tasks.

Identifying the required or appropriate project sequence is a strategic part of the project manager's plan for the execution of the project. In the early stages, or even later in very complex projects, it may only possible to identify the phase, eg, the construction phase, without being able to identify the details of the tasks. Identifying a phase is a useful step towards identifying the tasks.

3. Breaking the project sequence into manageable parts

The project life cycle helps to break up the project sequence into manageable parts. The phases usually include work of a related nature and the project manager can focus on this related type of work. So, in a design phase, for instance, one can focus on design issues and not have to deal with the day-to-day issues of the implementation phase. This reduction in the range of items to be considered at any one time makes the project sequence more manageable.

4. Promoting a sense of urgency

The project manager must have a strong desire to finish the task at hand, constantly strive to remove roadblocks, develop a clear understanding of what to do next and have the determination to get on with it. While it is recognised that this sense of urgency can go too far, it is important to maintain momentum and keep people feeling that work is being accomplished. The life cycle, assisted by the networks, helps to maintain this sense of urgency. Making progress towards major milestones in the project life cycle stimulates project management staff.

5. Deciding the acquisition strategy

At the very beginning, the choice of the project life cycle brings the acquisition strategy into focus. Acquisition covers such issues as who will do what part of the project, whether it will be a supply only or a supply-and-install contract, whether it will be a turnkey project and so on. The acquisition strategy is about the relationships to be entered into to achieve the project. The project life cycle forces one to focus on these relationships and helps in the analysis of alternative strategies. In the early stages, the project life cycle allows one to focus on the overall relationships to be entered into; their detailed content can be developed later.

6. Identifying appropriate staff qualities

Projects require a range of behaviours. In the early phases of a project, one needs more open and flexible types of behaviour, while at the end, one needs more focus on completion and more resistance to behaviour that introduces new ideas. The project life cycle allows the project manager to consider some of the personal qualities staff might have and encourage them in particular aspects of their personality according to the phase of the project.

7. Integrating activities

Integrating project activities and tasks over time and place is a key activity of project management. The project life cycle helps the integration of tasks over time by allowing the project manager to focus on how the tasks come together.

8. Timing of decisions

The project life cycle helps to guide the timing of decisions in that it allows the project manager to identify the times at which certain aspects of the project should be finalised. In fact, the project life cycle begins to indicate when the decisions are required and where the data may come from.

Consider, for example, three phases in the production of a new product: the product design phase, the acquisition of the manufacturing process phase and the production phase. These three phases sit within an overall life cycle of new product development. The dilemma facing the project manager is that, on the one hand, since each phase has to be fully complete before the next starts, the finished product may arrive too late in the market, while, on the other hand, if the decision to order the manufacturing process is made too early—that is, well before the product design phase ends—the wrong process may be ordered, with consequent costs. The life cycle is a key input to the timing of this decision.

9. Guiding the levels of contingency

The level of contingency is a key indicator of the cost and time performance on a project. In various industries, indicative levels of contingency should be available at the beginning of each phase.

Project life cycle phases

As defined earlier, a life cycle phase is a group of time-ordered tasks, and is a subset of the whole project sequence. The extent of the tasks included in any one phase depends on how they relate together. The beginning and end of the phase is chosen by the project manager. In practice, the beginning of a phase starts at the end of the previous phase, so one only needs to know how to identify the end of a phase.

The end of a phase usually represents the achievement of some milestone or the completion of some identifiable work (note that while a milestone may be used to designate the end of a phase, a phase can have milestones within it). In effect, one expects an identifiable deliverable to be produced as a result of a phase; this deliverable can be a set of information, a decision, a physical entity, or any one of a whole range of items.

When choosing or deciding on the end of a phase, one might reflect on the needs of downstream customers, that is, the needs of those who take over the project for the next phase. So an end of a phase should represent a convenient handover point. In this way, the phases can be chosen so as to facilitate quality management.

Modern projects involve a range of industries: most have a primary technical component, a software component and an organisational change component. There are, in fact, parallel life cycles that must come together at particular points, so the phases must be compatible at these points. So the choice of the end of a phase is influenced by the need to marry a set of project life cycles together rather than just dealing with one cycle.

The ends of phases can also coincide with major authorisation points. These may be requirements within the client organisation or of external bodies, such as a foreign investment review board or a government drug approval agency.

There is a view that the duration of phases should be limited. It is recommended that phases be short, lasting months rather than years. Phases longer than a year probably need to be broken up into subphases in order to maintain some sense of progress.[29]

Phases are set up to assist in the control of a the project. They must therefore help to make project progress more visible and also make the participants more accountable.

Reviewing life cycle strategies

There are a number of overall approaches to the project life cycle. The following life cycle strategies should be evaluated in the feasibility phase. In many cases, however, by the time that phase commences, the issues raised here will have been decided.[30]

The monolithic approach

One can do the whole project in one cycle and take the project through it phase by phase; this is called the monolithic life cycle approach. In this method, one divides the project sequence up into a set of phases, eg, initiation, design/development, production planning, production and handover. One first of all does all the work of the initiation phase, then all the work of the design/development phase, all the work of the production planning, then the production, followed by handover. Buildings and large bulky projects usually follow this approach.

The incremental approach

The second overall approach is to take a discrete, stand-alone part of the project through all the phases, and then take another part through all the phases. This is called the incremental life cycle approach, with which it is possible to allow work on the parts to overlap. The project manager identifies a part of the project and first takes it through the initiation phase, then the design/development phase, then production planning, production and finally the handover phase. Having finished that part of the project, the project manager chooses another part which is then taken through the whole life cycle, and so on until the project is finished. An example of this approach might be the gradual development of managerial control systems leading to an overall management system; one might completely develop the accounting system, then completely develop the costing system, then develop the marketing system, etc.

The evolutionary approach

The third overall approach is to define the whole project objective in a limited form, then be prepared to abandon it quickly after completion and move on to a better version of the project. This is called the evolutionary life cycle approach. Here, the project manager defines a project in a limited sense, such as easily achieved objectives, takes it through all the phases and hands it to the client. Next, the project manager starts working on a better version—better than the first but a relatively easy step up from it. This version then goes through all the life cycle phases, with the earlier version being abandoned when its replacement is ready. An example of this might be the gradual development of a piece of software, progressing from the early crude version to more sophisticated ones. In fact, most of the economy is going through large evolutionary life cycles—cars, computers, televisions, almost anything one can mention.

The fast-track approach

Superficially, this looks like the incremental approach. However, it is different in one crucial way: that while the project is broken into parts, it is not broken into discrete, stand-alone parts—the parts are interdependent. Typically, one makes decisions about the details of one part before one has finalised later parts, where these later parts depend on the earlier parts. An example in car manufacture is the ordering and design of production plant before the car design is complete.

Evaluating each strategy

The monolithic tends to have legalistic agreements, is highly structured, requires that the outputs can be specified, requires milestones, has little innovation and has separate sign-offs for phases. It requires a stable environment, low-level urgency and a single user. Signs that it is going wrong are the repeating of phases and unacceptable output. Compared to the incremental and evolutionary approaches, the monolithic approach is more strongly geared to resist change. It is also suitable for the more reserved client while the incremental and evolutionary approaches call for client involvement.

The incremental approach places the project providers and the client in a partnership arrangement. The controls are designed before the details are in place and one has to look at technical dependencies. This strategy can meet some urgency requirements, and is suitable where there are multiple users, and where there is ambiguity and controversy. It is also suitable in a changing environment where a basic facility can be defined and expanded in steps, with each step giving additional capability. A life cycle adopting the incremental approach can have certain parts done by the evolutionary approach.

The evolutionary approach is to start with a prototype and produce it quickly using ready-made items. The first prototype is quickly replaced by the second. This approach responds to an innovative environment that changes regularly, and should be considered if the project time is short and if the project is urgent, has indefinite scope, has a vague structure where innovation is implied, and where there may be strong resistance to change. Of all the approaches, this one gives the quickest response; users can help in the development. It can involve buying as much as possible off the shelf with no feasibility study.

Project life cycle control framework

A control mechanism can be associated with the project life cycle. The mechanism described here can be applied quite generally, not only in the control of life cycles. In this context, however, control is concerned with the identification of where, how, and what the project was *expected* to be at this point in time; with the identification of where, how and what the project actually *is*; and with the identification of a need for correction (arising from the difference between what was expected and what actually exists).

The specific correction to be taken is not the issue here—that action will vary in form and urgency depending on the particular situation.

There is a need to have a clear, easily understood control framework within which to manage the expectations, the actual outcomes and the need for correction. Such a framework is necessary so that people can not only be assured that the project is proceeding appropriately but they can also be prepared for any changes in their obligations. It is necessary in order that the objectives of the project have some reasonable chance of being achieved and that the full resources available to the project are properly used. It is also necessary to protect the morale of those working on the project sequence, and to promote efficient decision-making. The reasons are myriad. Adequate control will reduce confusion and waste and promote efficiency and effectiveness.

In order for control to be effective, it must be taken on board, agreed to and become part of everyday practice. This principle must apply right through the project management structure. For everyone to become involved, they need to be able to use a framework, so it needs to be simple, relatively straightforward and have the support of all participants.

Money is not the only feature needing control—time obviously needs to be examined. Other aspects also need a control framework; examples are quality of work, the management information system, project information, staff morale, staff capability and subcontractor performance. The method described here can be applied to these and other aspects.

Applying the framework: a cyclic approach

The project sequence goes through phases, which vary from project to project. The control framework is applied to each phase in turn. Each phase must then transfer to the subsequent phase a definition of what that next phase must achieve. Since the first phase has no preceding phase, it must find its own objective. The application of the control framework to the phases is cyclic.

1. Define the objectives for one specific aspect (eg, time, in order to meet a work deadline).
2. Plan how to meet that objective (eg, draw up a critical path chart).
3. Take action to meet the objective (do what has been planned in Step 2).
4. Measure one's progress (eg, plot progress along the critical path).
5. Compare one's actual progress with that defined in Step 2 above.
6. Decide if corrective action is needed.
7. Take corrective action.
8. Go on to the next cycle, ie, start at Step 1.

This process is often presented as a four-step model. Better-known examples are OPIC (Objectives, Plan, Implement, Control) and PDCA (Plan, Do, Check, Act). What they all have in common is the cyclic factor and that adjustments are made as a result of feedback. It is a surprisingly powerful framework—its power becomes very apparent when there is confusion and someone actually starts to apply it. Because it demands continuous application, which requires the commitment of time and resources, it is often let slip in order to respond to more pressing matters; this, however, eventually leads to loss of control.

Multiple control cycles

There is not one but a whole range of parallel control cycles on a project. This arises because the technical work of the project sequence (eg, bringing about organisational change, shooting a film) needs to be managed, while in parallel a wide set of indicators are being used to measure certain aspects of progress or certain aspects of constraints, such as time and cost. Multiple control cycles therefore consist of one cycle concerned with the technology of the ongoing task and an associated range of other cycles concerned with indicators of progress. An example might be useful.

There is a job to be done, say, writing a computer program to control a particular machine. It has been envisaged that the program will be 2100 lines long; at a rate of 15 line per day average progress, it is estimated that the task will take 140 person days. Two people are set to work on it with completion expected in 70 days. There are now two quite separate issues to be dealt with: there is the actual writing of the program and there is the measurement of progress. The measurement of progress can take the form of many indicators or measurements—time, cost, documentation status, etc.

A control cycle can be applied to the ongoing technical work to be done, ie, to the programming, while another control cycle (in fact, a set of cycles) can be applied to the indicators. Changing the behaviour of the indicator action can only be taken at the level of the technical work; to improve time performance, one needs to act at the programming level. Little improvement is available at the level of the indicator (some improvements, however, are possible using propositions or understandings of the behaviour of indicators—the fast-track technique arises from an understanding of the behaviour of the time indicator). Any adjustment has to take place at the level of the technical progress of the project. Thus 'time control' in project management is essentially monitoring time, with adjustment only possible at the technical work level; this also holds true for cost control and quality control.

Control of the technical work

The technical work can be set out as a sequence of tasks on a critical path network. One can develop a set of objectives for technical performance, such as response times or ease of use (the objectives). Having developed these objectives, one can then draw up a plan of how they are to be achieved, perhaps with a logic flow diagram (planning). Then one can do some work on the programming (implementation). Next, one can check the software to see how it is performing against the objectives (monitoring), and decide whether corrective action is needed. Corrective action may mean changing the objectives or modifying the technical approach (adjustment). Control is usually defined as monitoring plus adjustment. Thus, a control cycle has been established to control the technical content of the work. If the project manager has learned the vocabulary (even without the meaning) he or she can participate quite effectively in managing the technology (management of technology will be discussed later).

The cycle is applied not once but many times. In the process of its application, the objectives are developed, the planning improves, implementation improves and the technical approaches also improve. There is an ongoing elaboration of the problem and its solution.

Indicators linked to technical progress

In parallel with the technical work, there is the control of a whole set of indicators, one of which might be time, a key issue in software development. Linking the technical progress to an indicator is crucial; for example, there needs to be a means of linking it to the time indicator. A common way of doing this is to nominate points or milestones at which components of work can be clearly seen to be complete. Another way is to measure the amount of work done as a percentage of all the work to be done: one can then track the progress of work done on a time-scale.

One point should be made quite clear—time is only an indicator. It is an externally imposed criterion which of itself does not directly help to get the project done. It is a pointer to progress; it is not in itself progress.

Indicators and management

While the measurement of time does not directly contribute to progress, it can however affect the management of that progress, and thus make an indirect contribution. The measurement of time in relation to the project may indicate the need for more or less resources; for instance, it may indicate a lack of productivity or it may point to a potential threat to the project. A decision on a matter like this will have an impact on the actual technical progress of the project. The measurement of time in relation to progress, while not directly contributing to finishing the work, can be very helpful in the management of the completion of the work. This is the value of the indicator—it indicates the need to make an adjustment to the way the technical work is handled. For similar reasons, accountants might not seem to add to production, but their input on a project (such as data indicating overspending is occurring) provides vital clues to the need for better management or different management. Though much of management information does not contribute *directly* to production, it usually contributes *indirectly.*

Stakeholder criteria

Indicators can also be significant measures of satisfaction to be achieved, measures of success desired. The technical level exists almost independently of the stakeholders; it is in many ways independent of the interests of the stakeholders. Building an opera house over 15 years as against 40 years (in eras of technological stability) does not matter much at the technical level, but it does matter from the point of view of those who have an interest in using its facilities. Indicators are ways of measuring the extent to which the project is likely to satisfy the stakeholders, ways that are not measured by aspects of the technical process alone.

Controlling parallel cycles

The control framework actually has more than one generic cycle operating in parallel and interacting with each other. One control cycle must deal with the technical sequence. Let us put two cycles together, say the project technical work control cycle and the time management control cycle, and use the eight-step cycle described earlier. The management of these two cycles is illustrated on the following page. The OPIC four-step model would have Step 1 under O (for objective), Step 2 under P (for planning), Step 3 under I (for implementation), and Steps 4, 5, 6, and 7 under C (for control).

The important difference between the set of steps for the control of the technical work (the technical sequence management) and the set of steps for the control of time (indicator management) is that technical specialists are primarily responsible for the first set whereas the project manager is primarily responsible for the second set. The difficulty for the project manager is that the success of the project is determined by the technical process management; the project manager's tools, in many respects, are to be found in the management of the indicator.

The control framework cycle and project phases

So far, the focus has been on the control of work within a phase or within a part of the project. Now the work of the phases needs to be linked. Part of the work of each phase, except for the first, is to define the objectives of the next phase. This is one way in which to keep the project together as a single effort, one way to allow the progressive elaboration of the project, and one way for the control systems to be passed on from one phase to the next.

Management of Technical Sequence	Management of Indicator, eg, Time
Step 1a. Define objectives for the project phase (eg, the maximum memory requirement of a subroutine).	Step 1b. Define objectives for the indicator (eg, the time to meet some project deadline).
Step 2a. Plan how to meet that objective (eg, the logical structuring of the subroutine).	Step 2b. Plan how to measure the indicator (eg, choose CPM).
Step 3a. Take action to meet the objective (ie, do what was planned in Step 2a).	Step 3b. Not applicable
Step 4a. Measure actual performance (eg, what is the maximum memory used).	Step 4b. Measure one's progress (eg, plot how far along one is on the critical path).
Step 5a. Compare with the objective (eg, compare the desired maximum with the actual maximum).	Step 5b. Compare actual progress with the plan defined in Step 2b above (eg, is the work on, ahead of or behind schedule?).
Step 6a. Decide if corrective action is needed.	Step 6b. Decide if corrective action is needed.
Step 7a. Take corrective action, if it is needed.	Step 7b. Take corrective action, if it is needed. Go to Step 7a and return to 8b.
Step 8a. Commence next cycle, ie, carry out Step 1a.	Step 8b. Commence next cycle, ie, carry out Step 1b.

In order to illustrate the relationships between the phases, the eight-step model or the four-step OPIC model described above will be modified to a four-step model: Objectives, Actions, Monitoring, and Adjustment (see figure 1 opposite). 'Control' is contained within Monitoring and Adjustment; 'planning' and 'implementation' are combined under 'Actions', while Objectives stays as a discrete step. The difference here is that the Objectives are not developed at the beginning of each phase but developed in the previous phase, within Actions of the previous phase.

The framework starts with a phase: here a five-phase model is used.

Feasibility
Concept
Documentation
Pre-implementation
Implementation

FIGURE 1
Project control cycle

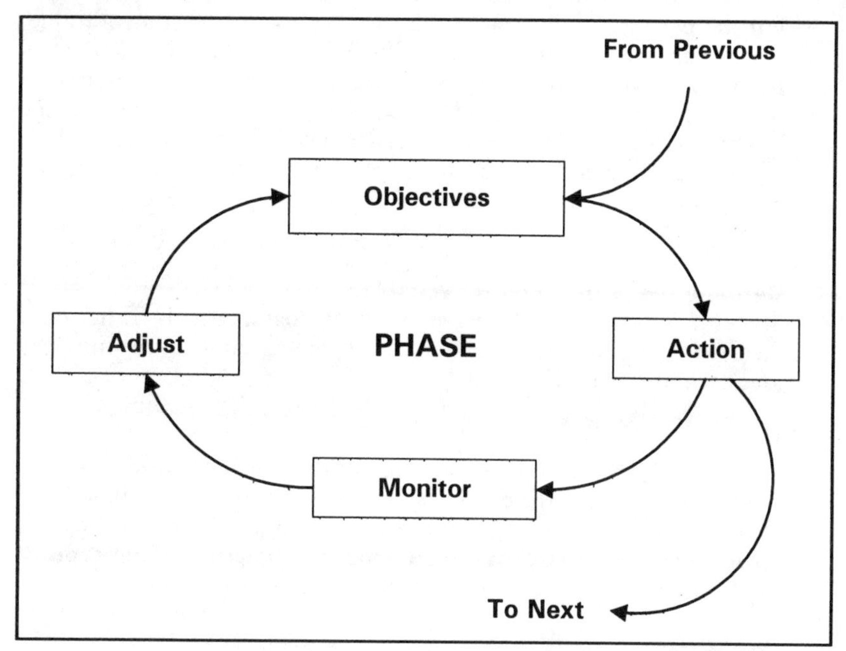

(Note that later we will introduce a Transition Phase as the first phase. For present purposes, this phase will be ignored: it does, in fact, establish the objectives for the feasibility phase. Also, later we will combine Concept and Documentation into one phase.)

The phase starts at the Objectives portion of the cycle during which the objectives for the phase is agreed (in most cases this will have been handed on from the previous phase). This is followed by the Actions portion during which various activities are carried out with the aim of achieving the objectives. One of these activities must be the development of the objectives for the next phase. During the Monitoring portion, the work is checked to see if things are progressing as they should. If not, the Adjustment portion is activated to bring things back into line; there may also be recognition that adjustment may in fact imply or force a revision of the objectives.

As the framework is applied to successive phases, the detail of the control gets finer and finer. As more and more information is gathered together, more and more specific control can be implemented. This closer control does not necessarily have to be applied from the top of the hierarchy; the more the management of the detail is effectively passed on down to operatives, the more efficient will be the process, with an accompanying increase in reliability.

The control framework cycle and cost control

For the purpose of illustration, the application of the control framework to project cost control will be illustrated. For each phase, a control framework cycle will be set up and the objectives will be inserted in the appropriate generic form. These objectives will be drawn from the Action portion of the previous phase. Monitor will be the generic checking to be completed, and Adjust will just be that, the word only. Before going to the framework in detail, some definitions are required.

Defining the project control variables

Six concepts are employed in project cost control—functional element, component, package, cost target, financial contingency and escalation fund.

Functional element

Functional elements are parts of the project which perform functions or some use, such as the cooling system in a car. Sometimes functional elements can contribute to a number of functions—a car body can be structural and also provide passenger protection. This type of multifunctionality can create difficulties in cost calculations.

Component

This is an identifiable part of a functional element. For example, a concrete wall is a component of a lift well; a car radio is a component of the car sound system. The same component type can be part of a number of different functional elements.

Package

A package is part of the project let under one contract or subcontract. Packages are chosen on the basis of the supply industry's method of delivery and other criteria. A functional element may be provided completely under one package but more commonly a number of packages contribute to a functional element, and a package may contribute to a number of functional elements. In software development, a package may be let out for the preparation of user manuals.

Cost Target

A cost target is an agreed amount of money within which the cost of a functional element or package is expected to be contained.

Financial Contingency

The financial contingency fund allows for errors in the estimates.

Escalation Fund

The escalation fund provides money to cope with the effects of inflation.

Moving through the project phases

The main purpose of the following is to show how the objectives are moved from one phase to the next (see figure 2 overleaf). The first four variables of those described above are used.

Feasibility phase

In the first phase of the project, one of the activities is to set the Overall Budget. Having set that, it is broken down into a set of allocations to functional requirements. Since it is quite a difficult process, these allocations are often not carried out.

What is being advocated here is that the project managers think about the way the money is spent across the key reasons for the project. The functional requirements here mean those aspects of the project that contribute to the functioning of the project. For example, in thinking about a cinema one might think about the division of the budget expenditure across the functions of car-parking, the foyer area, circulation, seating, projection and sound.

Assuming that these are the main functions one expects at a cinema, then it is necessary to allocate the funds across them. How much should be spent on car-parking and how much on the foyer? These are key questions that go to the core of the project's feasibility and success. It is important to look at the whole project and the complete experience of using the facility when allocating money. The reason behind this is that one will have to make trade-offs, eg, one might increase the parking but then one has to reduce the seating capacity. These trade-offs should be considered at the level of function.

Concept

In the concept phase, the technology of the componentry is identified At the technical level, the *functional elements* are converted to *components*. At the cost level, estimates of the costs of each are used. The componentry is not yet identified down to the proprietary level; this comes in the next phase, design development.

The objective is now to develop (in the form of component element cost targets) more detailed statements of the cost budget and still remain in the budget as set out in the Functional Element Cost Plan. During this phase, estimates of the costs of each of the component elements must be prepared.

The work of the phase includes the identification of the component elements of the project and the cost targets associated with them. In the monitor area, a check is made to ensure that the total budget is still on track. The Component Element Cost Targets are checked to ensure that their summations do not exceed the Functional Element Cost Plan. One can expect that there will be ups and downs, overs and unders.

FIGURE 2
Project control framework

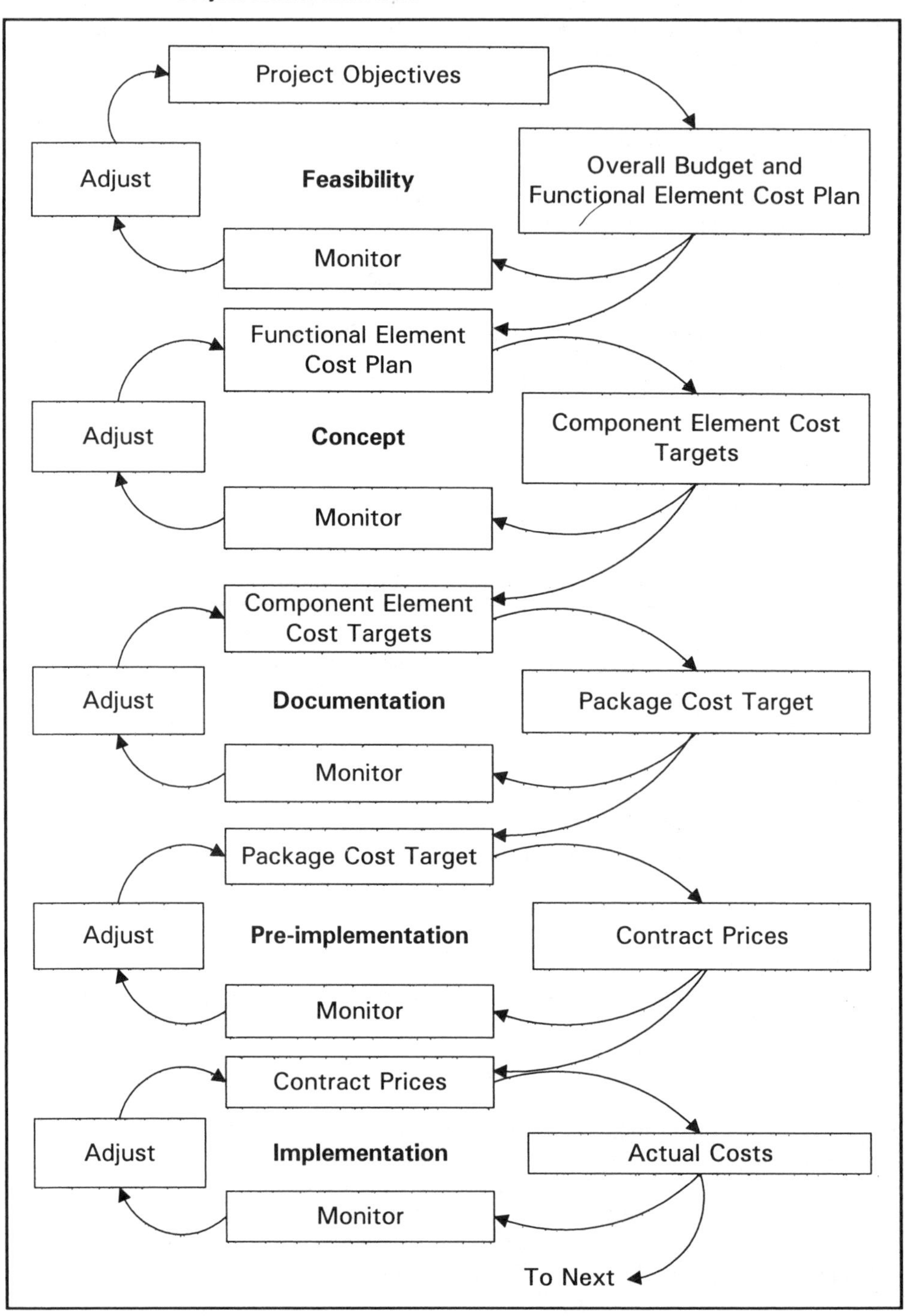

Most budgets are presented at this level, that is, at the level of component. The functional element break-up is rarely carried out (it is in fact quite difficult to get good data in this area) and the estimation process goes straight to the component element break-up. Sometimes it even skips this and goes to the package level.

Documentation

The Component Element Cost Targets are transferred to become the objectives for this phase. Now the technical work is to translate all the technical componentry, technologies and services into forms to which industry can respond with contractual offers. The component elements are brought together in packages, as appropriate. In this case, the technical work transforms the component elements requirements into packages that can be addressed by industry.

The Package Cost Targets are checked against the Component Element Cost Targets. The Package Cost Targets are now the keys to assessing the tender offers to be received. These become the objectives for the next phase, the pre-implementation phase.

Pre-implementation

The objectives of this phase are the Package Cost Targets. Part of pre-implementation is the work of inviting and assessing tenders and awarding contracts. The monitoring basis against which to compare the offers from industry is the Package Cost Target. Each target may or may not be correct and adjustments may have to be made. The contract prices offered by the successful tenderers are identified in this phase and passed on as objectives to the next. Note that here the project is not really under the control of management—it is now under the influence of the market. Of course, the prices offered can be rejected and the project abandoned.

Implementation

The objectives flow from the contract prices. Considerable administrative effort is now required to ensure that the payments conform with the tender documentation. Complications often arise because partial payments are made, instead of a one-off final payment. There are often arguments on the way—arguments about the quality of work or about the amount of work done. In this process of payment, the project manager needs to ensure that the money is available to meet the bills. On most projects, there are final payments to be made, but these are paid after the project is complete and contractual obligations have been met.

The control framework cycle and time control

The following table summarises the process described above for cost as it would apply to time and should be read horizontally, across the columns. The important thing to note is that the output from one phase becomes the input objective for the next. As the project proceeds, there is a build-up of greater detail in control of the time, and progressive elaboration. This goes hand-in-hand with the greater depth of detailed knowledge of the technical development of the project.

phase	objective	action output	monitor against	adjust
feasibility	overall time	network with low detail	overall time	adjust
design concept	network with low detail	CPM with more detail	network with low detail	adjust
design & documentation	CPM with more detail	CPM with detail program	CPM with more detail	adjust
contracts	CPM with detail program	contractual CPM	CPM with detail program	adjust
implementation	contractual CPM	weekly or fortnightly bar charts	contractual CPM	adjust

Adaptability of the framework

This framework can be applied to a range of indicators and to the main technical requirements. It is relatively simple and builds on what has already been done to develop phases. Objectives are passed from one phase to the next, which in turn develops the objectives to be passed on to the subsequent phase. The framework does require agreement on the phases, which may differ on different parts of the same project. It also requires that time and resources be devoted to reaching agreements and understandings, and to the maintenance of control data.

The project life cycle combined with the cybernetic control framework, although conceptually easy, is very powerful. The project manager needs to dedicate time and resources to the application of the control framework to the phases and within the phases. Simple as it is, it is a device that yields very powerful control over the project.

Part Two

The Project Life Cycle

Chapter Three

3

Transition Phase

This chapter is about the activities that are a preliminary to what is usually called the feasibility phase. These activities are not only necessary in order to initiate the project sequence; they have an enormous influence on what is and what is not in the project. Though most people see project management as starting at the feasibility phase, projects just don't come out of nowhere. It is activity outside project management that instigates the project; this is the transition phase, a phase that is more social and political than technical.

This activity goes on in the area of general management. A company does not suddenly launch a new product or abruptly announce an organisational restructure—before either event gets to the starting line, before the feasibility phase, something significant must happen. That significant something is agreement.

Key issues

Three fundamental questions need to be answered.

What is the project?
Who owns the project ?
Who pays for the next phase of the project sequence?

Oddly enough, the answer to the first question is not necessarily the most important; it can be answered in later phases. The second and third questions are more important, and the answers to them determine who controls the project. To execute a project, one needs to find someone willing to pay for it. The transition phase is about finding that person or entity. At this stage, it is not necessary to find someone willing to fund the whole project sequence, only someone willing to pay for the next phase of the project sequence, which is usually the feasibility phase.

The project advocate

The transition phase may start anywhere, any time and anyhow. It may initially be only a 'glint in the eye' or a certain disposition; it may simply pop up in conversation and get the response, 'That's a good idea'. It may only be a response to a perceived or suspected need; it may also be a carefully planned and considered option. A transition phase comes into being once someone decides there is a need or a want to be satisfied and that this need or want can be satisfied by a project (which may not yet be identified).

This person then starts working towards achieving the project. This person is the project advocate, and the transition phase is their domain.[31]

While an organisation will have many project advocates, only some will be successful. Typically, there will be many transition phases and many project advocates before a project is born. Many ideas will be advanced before one actually gets off the ground and progresses to the next phase of the project sequence.

Project advocates are usually competing with each other, either overtly or covertly, for the resources to get their project under way. They are also competing with existing processes in the organisation, processes that while due for replacement are trying to survive.

Where it fits in

It may be argued that the transition phase belongs to both general management and project management, and exclusively to neither. It clearly starts in the general management arena and then ends up in the project management arena. It is the sequence where one moves from considering a general trend of activity to working up a one-off change. So the following definition will be used as a rough working model.

The transition phase is that sequence of events that starts when the idea of the project or the identification of the need for the project occurs, and the sequence is directed towards the initiation of the project sequence.

This definition says that the transition phase starts when someone has decided to work towards getting or bringing about a project. They may have decided to start pushing for or working towards an organisational restructure, the revamping of a brand of shampoo, the introduction of a new car model, the opening of an overseas office or the formation of a business alliance with another company. Somebody has decided to go for a one-off.

It is also possible that the project advocate has simply decided that there is a problem and that a solution needs to be found. The decision to seek a solution is probably enough to start the transition phase (even if it does not strictly conform with the definition, that requires a recognition that a project is to follow). We have, of course, the possibility that the solution is not to be a project, in which case the transition phase ends.

For much of its duration, the transition phase is hidden and the project advocate may be the only person who knows of its existence. Later, when it is succeeding, it will have to emerge into the open where others can consider what is being proposed, and can support or attack it. Eventually, for the project sequence to commence, the proposal has to be agreed to.

The transition phase is not obviously or clearly linked to a particular project. It is very flexible and the project advocate may shift support from one proposal to another. In a way, one can only identify the transition phase when it is over and the project sequence has begun.

The phase itself is being resourced by entities who do not even recognise that they are funding the establishment of, or transition to, a project sequence. Much general management time is consumed in the transition phases of projects—executive time trying to identify how to deal with a range of problems or opportunities facing an organisation. From these deliberations, project sequences and projects will be born. This is an overhead on projects that is rarely seen and certainly not picked up by accounting systems. The cost of the transition phase is picked up within the realm of general management, not project management.

The transition phase can end in a number of ways. A successful end is the initiation of the work of acquiring or achieving a project. It is also possible for the phase to end in the abandonment of the project idea.

Project ownership

There are many meanings to or aspects of the word 'ownership'. It is a very subtle term, applying to those who are paying for the project, to those who will use the project and to those who have a stake in its execution or outcome. The following definition has limitations but it will be useful to us here.

Ownership is the belief that one's views in relation to the project should be given due weight—those with greater ownership believe their views should have greater weight.

The belief that one has ownership may have no legal basis, though it often has a political basis. Ownership implies that the views of a range of people need to be taken into account. Some of these views will, of course, have negligible influence but others will be of enormous importance. The project advocate will have to judge what is likely to be of most benefit to the project. To get the project sequence going, the project advocate will have to broker agreement among a sufficiently large proportion of this wide range of owners.

Ownership issues will turn up when a number of business executives are competing for a project, when divisions within an organisation are involved, and especially when a number of organisations are involved. Ownership will also be an issue in the external environment, which will be discussed later.

Ownership is a social concept, that is, it requires a society to make it work and it is dependent on that society for its existence. People agree, or are forced to agree, that certain ownership forms exist.

Ownership has an enormous impact on projects and their management; for project managers it has great significance because they will often have to work in situations where ownership is dispersed, diverse and unclear. As the project proceeds, ownership should become clearer. The project advocate needs to use the transition phase to sort out this issue. In fact, this dispersed ownership is sometimes the main underpinning to the function of project managers because they become the single point of control through which the decisions of owners are communicated and managed. Without the project manager, there is no single point for the management of the diverse interests; the ownership is dispersed. The project advocate, in the transition phase, has to identify the project manager.

From the project manager's point of view, ownership is the vehicle that permits the following: the licence to act, the licence to control the resources on the project, the licence to decide, the licence to sign contracts, and the licence to do many other things. In some cases, licences only come into existence because the project manager asserted them by taking action as if they existed.

A successful transition phase will identify the owners and ownership of the project and gain sufficient agreement for that ownership to be used to appoint the project manager and get the project sequence under way.

The project advocate needs to recognise that as the project comes to be accepted its ownership will disperse over a wider group of people. While some of this dispersion is essential for the project to proceed, it also means that others are taking over from the project advocate (which may be beneficial, but not for the project advocate).

(Later on, as others join the project, they will assume a level of ownership. While their sense of ownership will be confined to their part of the project, it also will be related to the overall project. Thus, as the project proceeds, there will be an expansion of ownership, both in quantity and diversity. Projects are one of the few aspects of modern life that can generate the 'We are all in this together' attitude, a very powerful tool of success. The project manager would do well to harness this sense of ownership during the later phases.)

Key players

To get the project under way, the project advocate has to find a number of key people. These are the client, the users, the project control group, the project manager(s), and another group that will vary from project to project but which will have an influence on, or the ability to influence, project outcomes.

The client

Someone has to pay for the project, and they have to pay for it during its creation, upfront, while the project sequence is under way. One might argue, for example, that customers eventually pay for production facilities; they may, in one sense, but they do so after the event, after the production facility has been built and paid for by someone else. That person, as far as project management is concerned, is the client.

The term 'pay for it' is probably too loose. On many projects, banks and financiers actually pay; they take a risk but take a second line of financial liability. The client takes the first-line financial liability (it is of course possible for a financier to take first-line financial liability, such as in a joint venture; in this case, the financier moves into the category of client).

The client is the entity assuming first-line responsibility for the financial liability for the bills associated with the achievement of the project.

(There are many service organisations that use the word 'client' but with quite a different meaning from that used here. They define the client as the person who is in receipt of the service, to whom the service is directly delivered, but who usually does not pay for it. These people are usually not powerful and are not regarded as such by the service deliverers. In project management, the client is powerful and significant, and is the entity that ultimately calls the tune—or should, if it is properly organised.)

Having just defined the client, it might be assumed that they can be easily identified. If the client is a single person who knows his or her mind, thén it is easy. But if the client is not a single individual, who the client is becomes less clear. In dealing with a couple, say a husband and wife, who is the client?

When entities having a legal existence, such as companies, are clients, their decisions can be formally recorded through appropriately executed documents. But many decisions made on a project will not be executed formally with documents signed, sealed and delivered; it would simply take too long. Such decisions will be made by people having appropriately delegated authority who go about making their decisions in a range of formal and idiosyncratic ways. In practical terms, they take on the role of the client.

Clients also include joint ventures. Again, while there will need to be a formalised way of decision-making, people act with delegated authority.

The users

A project arises out of a need identified by the ongoing business and will be handed over to the business once it is achieved. Someone or some entity, either singular or plural, will have to live with or take over the project and 'run with it'.

Of the groups of stakeholders in the project, one is the group that will take over the project when it is finished. For example, in the introduction of a new production facility, there must be a group who will run it when it is complete, who see it as in their interest to take over the project. If these people do not exist, the project should not continue. Thus it is necessary for the project advocate to ensure there are people who will eventually take over the project. An appropriate stakeholder will have to be sought, found and inducted into the project. Otherwise, one is involved in a highly speculative venture.

Clearly, some of these users should be represented during the project sequence so they can contribute their skills to it. The project needs their representation in order to develop their ownership of the project and thereby increase the chances of success.

The stakeholders

There is also a need to identify another group or groups associated with the project. These people have some interest in the project. Stakeholders are able to give expression and effect to their views on the project, and can be either supporters or opponents of it—perhaps supporters if their views are accommodated, opponents if they are discounted. Stakeholder groups can be quite large, varied and with widely differing views. (Note that a wider group of stakeholders will emerge as the project proceeds. The project manager, rather than the project advocate, will have to deal with them. Also note that users are included in the stakeholders but stakeholders covers a wider group than the users.)

The client project manager

Almost every project involves a range of project managers but only one needs to be found in the transition phase, the client project manager.

The client project manager is employed to advise the client and act on the client's behalf in the execution of the project. There is a professional responsibility to act in the client's interest. Acting in the client's interest is not a simple matter, particularly when a complex client and a complex project are involved.

Just because the client has the money, it does not necessarily follow that the client is a strong manager. Projects can hold dangers for clients, for all the usual reasons such as choosing the wrong project or the expected demand does not emerge. There are also dangers that have to do with the execution of the project and the management of contractual relations. Because the client can unwittingly be led into a project which is expanding out of control, the client must actively participate to protect its interests. What is very important is that the client recognises the need for a project manager.

In exercising their role, project managers take on a form of agency—they make decisions on behalf of someone else and commit for someone else. That commitment is often contractual. The legal position of project managers is gaining attention, particularly in relation to their professional liability. Clearly, the law of agency plays a role in the legal framework within which projects managers act.

The client project manager does not have to come from the client's organisation. There are many management consultant companies which provide project management services to clients and act in the role of client project manager.

Project control group

A device often employed where the client is an organisation, a joint venture or any entity more complex than one individual, or where many views need to be taken into account, is a project control group (PCG). The PCG is a committee that may have a range of representation; typically, it includes a representative from each section of the client organisation that needs to have a say. It may include a stakeholder (but this is uncommon), and while it may not include the client project manager, it usually does. The project advocate (or later, the client project manager) should make it his or her business to be part of the PCG or at least have the right to attend all meetings and speak.

Ideally, the emergence of the PCG should occur during the transition phase—bringing it into existence will certainly enhance the project advocate's position. However, given the political nature of the transition phase, that may be difficult, and the PCG may be formed at a later stage in the project, but hopefully at not too late a stage.

The PCG is delegated either the overall decision-making authority in relation to the project, or a constrained level of authority. The constrained level of authority can be arrived at by various techniques such as a dollar limit, a percentage limit, or the sum of the delegations of the individual participants. Whichever method is used, many PCGs find ways to broaden their decision-making ability, such as by breaking down decisions so that each component falls individually within the delegated authority.

The PCG's authority to act needs to be defined. It would usually have authority over most day-to-day operations. Its role can be clarified as the project moves on; initially, the PCG may have quite limited financial authority which can be expanded later. Some PCGs exist only to provide recommendations, but this can be quite cumbersome.

The PCG is given the job of balancing the competing interests involved in the project and of reaching agreements that are viable in relation to the project. It is a mechanism that can accommodate the ongoing decision needs of the project with the need to balance differing views. There is, however, a need for the PCG to maintain a pace of decision-making that is consistent with project requirements, particularly in relation to change management, responding to contractual conditions that have time limits, and the letting of contracts.

Because in any reasonably substantial organisation, the interests of the client are difficult to fathom, the members of the PCG, employees of the client, may be more accurately described as stakeholders. This would be the case where the company or organisation is acquiring something of benefit to a range of interests in the company, such as a new computer system. An exact definition of the client's interest is difficult here and the members of the PCG may be appointed explicitly to represent the diverse interests. Thus the PCG can represent quite different aspects of ownership. In so far as the members of the PCG act in the interests of parts of the client, they may be regarded as stakeholders; in so far as they act in the interests of the client entity, they may be seen as client representatives.

A PCG may be seen as something similar to a board of directors, but there are quite important differences, a major one being in the area of decision-making. Boards of directors are supposed to represent the shareholders, to represent their best interests. Thus they take on a corporate perspective in relation to the company activity and their decision-making is supposed to represent the common good of the shareholders. It is doubtful if a PCG would be seen as taking this corporate view; its members are usually seen as representing and promoting special interests.

Agreement and support for the project

The transition phase is both political and missionary in nature. It is political in that it is about deciding who has control over the disposition of significant resources, and it is missionary in that it is about persuading others to support the project, which at this stage may only be a vision or a possibility.

Building up commitments

The crucial activity during the transition phase is one of building up commitments using a step-by-step process, going from a weak commitment, such as the nod of a head, to a stronger commitment, a handshake, to an even stronger commitment, a signature, to an even stronger commitment of agreement to fund the project or to fund the next phase. The sequence is not a one-step process but one of iterating back and forth, massaging the definition of the project, taking account of interests and building up the level of expressions of commitment, all leading to a clarification and agreement as to who will pay for what, who will provide political and moral support, and agreeing to oppose competing project proposals.

Note that the idea is one of gradually achieving agreement, not suddenly arriving at a 'yes/no' decision point. People familiar with the selling process will be aware of the need to take prospects through various stages before attempting a close, attempting to gain agreement. In fact, one of the major agreements that salespeople strive for is the agreement to continue discussing the sale. It will be a similar situation for the project advocate: throughout negotiations, one of the key agreements to be sought is the agreement to continue discussing the project proposal.

Negotiating political interests

The transition phase is political because, before a project is launched, there needs to be agreement on the expenditure of funds and resources. With widely differing needs competing for limited funds, and with many options available to satisfy those needs, reaching agreement on what should actually be done can be difficult. Projects usually require large resources. Except for very small projects, resources can only be brought together by agreement and through some form of co-operation (which may involve considerable pressure). It is crucial that the project advocate recognises this.

This phase is political because it must take account of power—identifying the powerful players and assessing the power of participants is a useful and essential activity. By extension, this phase is political because it must also take account of a range of interests. Projects seek to meet the needs of people—they do not exist in an emotional vacuum. The interests may be personal or organisational, but more commonly they are a mixture of both. The project advocate needs to build up a picture of the interests to be served and how they will be served; this involves a balancing act. The project advocate, more than likely, will have to choose which interests are to be served and which are to be opposed, ie, make political choices. These choices will not always sit nicely together and will sometimes have moral elements.

As the sequence progresses, people will develop loyalties. In some cases, they will simply be personal loyalties, but in others, a tribal loyalty will be involved. Project advocates must not only pick their way through the 'legitimate' needs, the interests of the powerful, but also through the minefield of loyalties. To add to the project advocates' difficulties are the demands that will be made on those who already support the project; they will have to face calls competing for their support—the project advocate will be only one such caller.

As the project sequence proceeds, sources of potential disagreement lessen (though regrets may continue). As decisions are made and executed, the range of project options is reduced; possibilities are closed off and flexibility of action is further restricted. Thus the political element will decline—at least, that part that is concerned with the form and nature of the project. There will be a need for continuing political activity to ensure ongoing support. Later still, the political activity will focus on dividing up the benefits of the project.

Taking the missionary approach

In the problem-solving approach, one knows what one wants to do and focuses on how to achieve it. In the missionary approach, one is essentially promoting a solution or promoting the need for a solution. 'Missionary' implies vision and the need to persuade people to subscribe to that vision. Much of this, however, will not be rational.

The transition phase requires the missionary approach. The project advocate needs to take a stance promoting some vision and try to gain sufficient support for it. The vision is that something is required, even though the project advocate may or may know what is wanted or who wants it.

Later in the life of the project sequence, the missionary approach will diminish and the problem-solving approach will come to dominate. Others will emerge to solve the many problems involved in trying to execute the vision; these people are commonly known as project managers.

Reaching internal agreement

The agreement being sought during the transition phase concerns who pays, and for what, and who controls the execution of the project. This internal agreement is among the project participants, and differs from those agreements that are yet to be negotiated with external bodies, for example, with environmental lobbyists or with other participants in the product distribution chain. Internal agreements are needed to get the project started. To say that they are internal means that they involve only the project proponents, that is, those who see themselves as promoting and executing the project. Agreement has to be reached on who is participating and on what terms. While the project advocate is working in an internal environment, this internal environment may cross organisational boundaries.

Projects emerge from the internal inter- and intra-organisational political activities that form part of the stuff of organisational or company politics. These activities will usually be concerned with finding ways of satisfying wants and needs. In the process, they will be concerned with resolving conflicting interests and with building up a coalition of support for the project, while at the same time overcoming coalitions opposing the project.

Handling controlling cliques

At the heart of practically all organisations such as companies or societies there is a group of people, usually quite small, that makes the decisions, even though there may be formal processes that give the impression that decision-making is broadly based.

This fact should not come as a surprise. In democratic systems, where formal decisions are made in a parliament, senate or similar chamber, the real decision-makers are the party leaders whose decisions are then essentially rubber-stamped by the formal legislative body. It is the same in gaining support for a project. Projects requiring internal political support need the approval of that small group of decision-makers, the controlling clique, and it is to them that the project advocate needs to direct attention.

The existence of these controlling cliques almost seems technically necessary; they allow negotiations to take place and support to be efficiently gathered. Without them, decision-making would be very complex if not impossible.

But these cliques are not all-powerful. Members of these small groups see themselves as operating in an environment which has both threats and opportunities, and they are mindful of the views of their support base and others. While company executives have the company's shareholders to consider, they will also be mindful of the views of their immediate superiors and other executives who might be in competition with them. Thus the project advocate has to analyse and know how the project proposal meets the interests of the members of the controlling clique. This means that the project advocate has to ask questions.

Some organisations may not be under the secure control or influence of one small group; there may be a number of groups competing for control. In this case, the project advocate has to either predict which group will ultimately have control and deal with that group, or deal with all the contenders. If the project is not part of whatever divides the factions, if its existence is either advantageous equally to all or is of no significance to them, then negotiating with each faction separately should not cause trouble. Where the project is part of the dividing line between the factions, then the project advocate may have to decide which side to support, or abandon the project. The project advocate should not end up with a project having two or more clients.

Focusing support and opposition

In the absence of a controlling clique, it may be necessary to actually set one up in order to focus support or opposition to a project. Where support/opposition is scattered and where that support/opposition could influence the project outcome, unless one group can control it there can be no real political progress. To set up such a group may be politically necessary.

When promoting a project or trying to ensure continuing support, the need for a small group to exist to allow negotiations to take place should be recognised. But before setting up such a group, note the conditions: the support or opposition is dispersed and also has a significant influence on the project.

Identifying essential stakeholders

Some stakeholders and their representatives have to exist, for example, the user groups referred to earlier or financial stakeholders. If they did not exist, the successful outcome to the project sequence would be threatened. The project advocate needs to identify the essential stakeholders and their representatives in order to maximise the project's chances of success.

Giving the project the cloak of morality

Morality is about values. It may be personal, organisational, or both, and, as with interests, personal morality and organisational morality may clash.

As a political rule, the project advocate should note that all action (including project proposals) needs to be given a cloak of morality.[32]

A cloak of morality reduces the need for the use of power (a more fundamental political rule is that the explicit use of power should be minimised).[33] It is claimed that the project is, in some way, for the general good or can be justified by reference to some objective criteria independent of the interests of the people promoting the project. It could be claimed, for example, that it adds to the quality of life, it is for the good of shareholders, it improves efficiency, it recognises the higher level of risk taken by some, and so on. A project can be given many cloaks of morality but it must have at least one.

In internal discussions, moral issues may be put in indirect ways. For example, in a dispute about a particular proposed project, there may be discussion of the risks associated with it—a seemingly innocuous issue, not involving morality, and which even has a whole body of mathematics devoted to its understanding. An opponent of the project could take a moral stand by attacking the project as 'a reckless use of shareholders' funds'. 'Reckless' implies irresponsible behaviour and not giving due weight to the interests of others; it implies a negative, undesirable value. Calling it reckless would be quite different from labelling it as a 'high risk' or 'brave' project which, after due consideration, might be accepted or rejected.

Most project managers see themselves as honest and honourable. Many of them, while enjoying the implementation phases of projects, find political activity unattractive and, in some cases, regard it as bordering on the immoral. They essentially view political operators (among whom are project advocates) and their activities as undesirable. They sometimes react by not becoming involved in a project until agreement is reached—a kind of 'Call me when you know what you want' attitude. There are many reasons to hold back when new projects are being mooted, but it is important to recognise that starting a project, getting it up and running, and building up a head of steam requires political activity to gain agreement. That activity involves dealing with interests and morality. The project advocate needs to be sensitive to interests, particularly the powerful ones, and might need some moral agility.

From vision to project

At this stage, nothing of the project exists—it is only a vision, maybe a 'bit of blue sky'. The vision is malleable, it can be altered with just a few words. Here the concern is with the elasticity of the project in terms of its reach, size, scope and form.

Taking cost as a measure of the range of possibilities, project proposals can swing in cost by factors of ten or more. An oil-change system can be an oil drum with a portable pump (costing say $4000) or a complete oil-reprocessing system with storage plant and fixed pumping (costing say $400,000). The addition of one word or a change in a number can radically alter the proposal. Another example is where a vague view of the likely demand for cinema seats can change the size of a cinema from a few hundred seats to several hundred or maybe thousands of seats.

Thus, in the transition phase, the range of issues over which there can be disagreement is huge. This has very important implications for the tactical approach of the project advocate. Big, exciting proposals, have a natural attraction but can frighten the safety conscious; modest project proposals are duller, may satisfy the safety conscious but bore those with a stronger sense of adventure.

The project advocate is faced with an overwhelming range of possibilities both in size and form: the big challenge is how to present the project. With so many competing issues to consider, the presentation will have to be adjusted to suit the circumstances. The project advocate must focus on two things: the clarity of the proposal and its malleability.

Because there is no simple connection between wants or needs and the means of satisfying them (ie, the form of the project), different desires can lead to the same project while the same desire can lead to a range of projects. The desire for entertainment, the desire to have a win in an election, and the desire for an image of cultural sophistication may lead to an opera house project. But the desire for cultural sophistication may also lead to a concert hall, a theatre, an art gallery or a museum. Many wants and needs can find their expression in the same project while the same want or need can lead to support for a wide range of projects.

Striking the right balance
At this early stage, the project is least clear. It is undefined and will change its form and size as negotiations and discussions proceed, and here, the problem centres on how quickly or easily that range in form or size should be reduced. The main tactical issue relates to the difficulty of maintaining support for a vague proposal, while the needs of effectiveness demand keeping options open to give the best chance of finding a solution.

Because there is such a wide range of options or possibilities, there is a strong tendency to reduce that range. There at least three reasons for this.

1. It can arise because a reduction in the number of options means a reduction in confusion. When only a limited number of options is discussed, the people involved have a better chance of understanding them and following the arguments for and against them. Reducing confusion prevents competitors with more specific project offerings from distracting supporters

2. A wide range of open options leaves many people, probably most, feeling very uneasy. Only a small number of people will be happy to discuss a project that only has a vague outline; others will want a much more precise definition. A 'culture centre' may be enough for some, but others may want it further defined as an 'opera house' with descriptions of shapes and sizes. Thus the depth of definition of the project depends on those involved. The need for this specificity is a personality trait—there are those who like to keep their options open right up to the end, while others like to get things tied down as soon as possible; there are those who like to play with the possibilities and those who like to close off the details. Most people prefer clarity and specifics.[34]

3. Perhaps the most important aspect is that reducing the range of options helps focus support. People find it difficult to support and remain loyal to the non-specific; gaining support for some nebulous idea is more difficult than gaining support for a specific action. The specific action of cleaning up the world on a particular day has gained widespread support; it is doubtful if support could be gained for action on a day that varied from person to person.

Just to make it a little more complex, in the process of gaining agreement, the project advocate is trying to assemble a coalition of support. A single proposal may be sufficient to attract a strong enough coalition but in other cases might not. One proposal will gain the support of one group while another will gain the support of another group. A proposal where an opera house is the outcome will get support from one group, while another proposal where the outcome is a concert hall will get support from another group. Thus the coalition of support for the project varies with the definition of the project. In the case of the Sydney Opera House, both coalitions of opera and concert were addressed (not without later tensions and costs!). The project advocate needs to balance the definition of the project with the coalition of support to be achieved.

There is a tension between clarifying the idea so as to gain support, and deliberately being vague or leaving options open in the hope that thereby a better project will result. The need to clarify the project at an early stage is probably against the interests of achieving a good project outcome; achieving a good project outcome requires that options should be kept open for as long as possible, waiting for the best option to emerge after appropriate examination and testing. Therefore the project advocate needs to be sufficiently focused to allow people to give their support, while keeping things sufficiently open to allow better options to emerge.

Concentrating on needs and wants

Which all takes us back one step further, to the question as to why the project is needed or wanted. The project advocate can probably improve the tactical position by developing a strong definition of specific wants or needs to be met—rather than specific projects—but having also a specific project proposal to present as a possibility, to launch as a distraction, as a kite or as a simple holding exercise. If agreement on the specific wants or needs looks likely, then the kite may be used in the political battle against competing project proposals, by attacking them on the basis of their failure to meet the agreed wants or needs. If this tactic can be followed, ie, concentrating on needs and wants, it keeps options open and makes the right choice of project more likely (assuming, of course, that one has identified the right needs and wants).

Effectiveness versus efficiency

Because this is a project, because it is a one-off and because project efficiency is hard, if not impossible, to measure, effectiveness is paramount. The question 'Will it do the job?' is infinitely more important than 'Will it do the job well?' or 'Is this the cheapest way to achieve the project's objective?'.

The last two questions are of course very important when the resources involved are large, but the reader is warned that logic and politics demand that the first question about effectiveness takes precedence. The inefficiencies will be seen later, but by then it's too late—people will have moved on, and any loss of opportunity will be dealt with by the historians. Project advocates need to keep the issue of effectiveness clearly in focus and not be distracted by the massive detail and difficulties associated with efficiency issues.

This is not to imply that cost is of no concern; of course it is. But cost is of concern in terms of ensuring that the cost limits are known and that the income from the project will exceed the costs. That there is a profit is crucial, but profit maximisation is a bonus, if it happens. The project advocate argues on the basis of effectiveness.

In summary, the project advocate has to find or establish a client, find the users and identify the key stakeholders; set in train a process to identify and appoint the project manager or managers; see to it that the project control group is established; and concentrate on effectiveness. Throughout this, the project advocate has to adopt a missionary and political approach aimed at establishing an effective ownership structure.

Brief for the feasibility phase

The key document produced by the project advocate is the brief for the feasibility phase activity. This document needs to be agreed with the new 'owners' of the project and must specify the needs and wants to be met, along with other agreements that have been reached, such as the general form of the project.

However, it can happen that neither the project objectives nor the set of tasks is known. In this case, the project advocate can only set milestones; these may take a number of forms but at the minimum they will have a date deadline and/or an amount of money or resources allocated. This happens implicitly on all projects, at least in the very beginning. These milestones might be expressed as, 'Let's see how much we know in two weeks' time' or 'Keep on examining the issue and we will review it when $20,000 has been spent'.

The other key output is the resourcing, selection and appointment of the staff or consultants to carry out the work of the feasibility phase. Some of this work can be done internally, by employees of the client, but often outside people will be needed. To do this, the project advocate must know where to go to find the right skills. In a very difficult situation where the correct skills are difficult to identify, the project advocate needs to recommend a set of steps leading to the appointment and briefing of the people who will carry out the feasibility phase.

We are now at the point where most people think the project starts.

Transition phase: inputs and outputs

INPUT
Almost anything from anywhere—dreams, desires, personal and various other agendas, market conditions, internal pressures, external pressures, strategic plans, external aggression, etc. Neither the objectives nor the methods may be known.

↓

PROCESSES
Sort out control of the project: decide who owns it and gain evidence or recognition for that owner.
Define the project to meet as wide a range of needs as possible.
Gain agreement among at least some stakeholders.
Gain financial commitment, at least for the feasibility phase.
Elaborate the needs to be met.
Open up possibilities for evaluation (the definition of the possibilities may be constrained to suit purposes).
Provide the basis for the next phase.
Examine time and cost.
Decide the next milestone.
Select the team to carry out the feasibility study.
Prepare brief for the feasibility team.
Deal with perceptions of risk.
Champion the project.

↓

OUTPUT
Recognition of the owner. Brief for feasibility team which includes:
Problem or project outline.
Objectives of stakeholders.
Stated political considerations.
Expected deliverable from the feasibility phase.
Constraints.
Statement of assumptions.
Budget and timeframe for the operation of the feasibility team.
Possibly a budget and timeframe for the project.

Chapter Four

4

Feasibility Phase

The feasibility phase is the sequence of tasks established specifically to justify the acquisition of the project and decide how it will be acquired. To do this, the definition of the project must be sufficiently developed to allow the project to be examined against a range of criteria and to allow people to make decisions on what arrangements need to be made to bring the project into being.

Assuming the work of the transition phase has been carried out properly, there is a client, a set of users, some key stakeholders, a project control group and a project manager. If any of these is missing and is considered necessary, then that part of the work of the transition phase needs to be completed.

The client is required to resource the feasibility phase. This means providing funds and facilities. The feasibility phase resourcing needs to be treated in the same way as other project sequence resourcing: budgets have to be drawn up and time estimated. The more important issue is probably deciding how much money to spend on the phase. Having decided the amount of money, it needs to be translated into human and physical resources. The client should use the project manager in identifying the resource requirements.

Three main pieces of work need to be accomplished in the feasibility phase: further clarification of the project; the testing of the viability and suitability of the project; and deciding how to acquire the project. As can be easily imagined, these three main pieces of work do not stand totally alone, but interact. As always on projects, one will end up cycling or moving from one piece of work to the other in order to maintain progress with the practical work of the phase.

At the end of the feasibility phase, a very important baseline document will be produced; here it is referred to as the project proposal. The project proposal outlines the purpose of the project, describes how it is to be achieved, where the finance will come from, how the project will be managed and administered, the type of contracts that will be used, and any legal issues that have to be managed such as environment, health and safety. It also covers assumptions that have been made, the project timeframe, cost parameters, procurement strategy, and so on. It is a document that relies on complex work and reasoning in its compilation, but in its final form at the end of the phase, should be a clear, precise, action-orientated document.

Project clarification

The project may be clarified by means of a written document describing the project in more detail. If the project is a physical entity, then sketches and drawings may also be used. To start, we assume that there is a general idea of the project or a definition of a problem to be solved. This should have been sorted out in the transition phase; if not, it needs to be done now.

1. Identifying the main elements

As feasibility begins, there is practically no detailed data on the project. The description of it is at best vague—all that is known is that the project may be something to be made or built, or a response to a problem which has just been identified. Its size, scope and reach are not yet clear. It is necessary to clarify the project to the extent where one can make an informed decision on whether or not the project is justified.

By way of example, with a cinema complex, one may simply know that one wants to build a theatre complex—one does not even know just how many seats are to be in the theatre. In relation to a piece of software about to be developed, one has simply decided to develop it to solve a particular problem; one has yet to make many decisions about the required functionality. In relation to a business, one may have only decided that changes to the existing product line are needed in order to keep market share; all that has been done is to identify a declining market share as a problem.

At this stage, however, we do not want to completely specify the project. We need to decide the size of the cinema complex, its parking requirements and a few other things, but we don't need to produce the working drawings. We need to list the functions to be included in the software but we don't need to produce the code or detailed specifications. We need to identify the new product or the new marketing strategy but we don't yet need to plan the detail of the advertising campaign. In feasibility, all of these decisions are to be taken to a level of detail sufficient for us to decide whether or not the resources should be spent on bringing the project into being or, at a minimum, just taking the project to its next stage of development.

The extent of clarification required varies from industry to industry and from company to company. The more experience a company has, the less the need for clarification, but the practice will be the opposite!

2. Reviewing decisions

The steps taken during the feasibility phase to clarify the project have an enormous impact. While they have less impact than the steps taken in the transition phase, they will have more impact than subsequent phases. It is widely believed that these early phases have the biggest impact on projects. While no proof for this belief exists, it has an intuitive feeling of truth about it; in practice, during the early phases, one will find that just one or two words will bring big changes to the project.

Now is the time to carefully review the work to date and to seek the better project, to make a good choice. Given the importance of decisions at this stage, it is surprising how difficult it is to get people to formally and properly examine the status of the project, to put energy and resources into finding a good solution.

There will be pressure to get on with the project, to make decisions and to close off the sequence of choosing the project. The best way to resist this is to propose and manage a formalised sequence of activity which can be dressed up as much as possible as a usual and necessary project management activity, all the while continuing the search for answers.

3. Solving problems by design

To improve the definition of the project, the project manager needs to gather data and introduce some imagination or innovation into the sequence. The arrival of imagination is not that easy to guarantee, but by creating the right environment for it, it will turn up surprisingly often. It will require some of the design skills normally associated with the next phase, the planning and design phase. While the process is quite common, it is worthwhile looking at one model of how design decisions come about so that one can have some confidence in managing the sequence.

The important point here is for the project manager to be aware, to keep an eye open for the imaginative idea and to recognise it when it turns up. Often it will turn up in another context, when everyone's concentrating on something else. It can come from anywhere, so be ready to grab it.

Design is equated with artistic processes and thus given a mystic label. This is fair enough but it should not prevent lesser mortals from trying their hand at it and trying to improve upon things. The design process does not only apply to built or manufactured objects; there is an element of design in all problem-solving processes. In fact, the phrase 'problem-solving process' sometimes can be substituted for 'design process'. A possible design process consists of the following sequence of steps:

- identifying a user
- defining a need to be met and criteria to be satisfied
- proposing a way to meet that need
- checking the proposal
- modifying and improving the proposal
- going back to proposing a way to meet the need until a few viable possibilities have been found
- choosing one of the viable proposals

There are a number of things to note about this process. First, the process of identifying the need is a reasonably logical process requiring some sensitivity and the kind of questioning skills commonly developed in the training of salespeople. Defining the right problem is of vital importance—there are few things more frustrating than spending resources on a useless or wrong project.

Second, the step of proposing a way to meet the need is not a logical process—it is a jump, a leap. All identification of solutions invariably involves a jump; having jumped and found something, one can then fine-tune it in the process of evaluation. The jump is that part of the process most likely to provide the innovation or the imaginative solution. These jumps, or imaginative leaps, can be assisted using techniques that allow a free flow of ideas, such as brainstorming. There are other, newer, ideas on how one examines possibilities, such as 'Seven New Tools' of quality.[35]

Third, one finally has to make a choice. At the end, there will usually be two or three options with no clear winner. In the final analysis, one will simply have to say 'I prefer this one'. The usual step is to choose one and recommend it to the client; the client then accepts or rejects the recommendation.

In making the recommendation, one might feel constrained by some need to identify the optimum. The concept of the optimum is a nice mathematical device but of little real use in actual live management situations. One only knows what the optimum might have been long after one has made the choice (sometimes one will never know). In other words, the right decision becomes obvious after the event—which from any manager's point of view is too late. One makes choices recognising that there is always risk. Projects that suffice are recommended here: projects that meet the objectives and the criteria but will not expose the client to severe or catastrophic risk. The reader may have a different view.

This design process is a very useful problem-solving process and can be used in many situations. It will be used again in later phases.

4. Defining needs and criteria

Whether or not one needs to distinguish between needs and wants will vary from situation to situation. Here, the distinction between them will be ignored, with both coming under the label of needs. A need is a desire to be satisfied. It always relates to people or to organisations or other groupings of people. In projects, these people come under the headings of the client, the stakeholders, those who produce the project and the users; sometimes there are the consumers. These categories may overlap or turn up in different forms. However, it is these people whose needs must be considered.

Needs can be associated with criteria, conditions to be satisfied by whatever is proposed. Criteria can include issues of cost, time, public acceptability and risk. For example, a need to regain market share may lead to criteria that emphasise the reuse of existing equipment, a timeframe and a capital cost limit. The criteria may specify that it is to be on the market within six months, that it is to cost less than a nominated amount to manufacture and so on. Criteria can be non-numerical and non-quantitative; they can also be quite subjective such as 'The product shall look attractive'.

For some, the distinction between needs on the one hand and criteria on the other serves little purpose, but it should at least be given some consideration. By jumping ahead into an explicit solution, real opportunities can be lost or quite silly items chosen. What one will find is that projects have to satisfy both needs and criteria.

When there are too many needs to deal with, priorities have to be set. Activities such as market research, gathering statistics on factors that influence people, trying to establish magnitudes of demand, soliciting views, examining technological forecasts, checking some technical viability and so on, help to both identify needs and develop criteria. Some of this data is gathered within the context of justifying the project but it is also very useful in providing an input to the design process or problem-solving. One does not need very precise data, one simply needs order of magnitude data. The magnitude of things (such as the size of the pocket computer) and the extent of what is being considered (a one-branch reorganisation as against a whole-company reorganisation) have a big impact on the form of the project.

5. Identifying the extent of the project

It should not be necessary to say that one needs to recognise the full extent of the project. However, it has to be said because writers in the area of project management identify failure to manage the entire project sequence as a considerable weakness in project management.[36]

That one should recognise the full extent of the project is often ignored, and very easily forgotten. There are a number of possible reasons for this. The first is that while the project will always extend beyond what appears to be the substantial work of acquiring the project, most people focus on just that part, the substantial work. For instance, a software package is much more than the writing of code, yet this will be seen as the only part of the work to be done, with the rest just detail.

The second reason, and probably of more significance from the project manager's point of view, is that no-one seems to have full authority to deal with the whole scope of the project. For instance, the introduction of a new product will touch practically all areas of a business, but the only person with this breadth of authority is the general manager, who is occupied with other things. Thus there is a power vacuum into which the project manager must step and assert authority. If the need exists, the project manager will usually find that others will consent to this move. This is a significant managerial action by the project manager.

It helps to identify the full extent of the project if it is recognised that most projects have both technical and social components—the bit that is often ignored or missed is the social component.

Technical Component

The technical component usually involves more than one technology and almost invariably involves information processing technology. A new product will have all the technology of the function of the product but also the technology of its manufacture, storage and delivery. A new building has all the technology of traditional structures such as concrete, steel, brickwork and timber, and it also has the more recent technologies of air-conditioning and lifts. Not only that, but buildings now have a whole range of communication technologies, technologies beyond the electrical wire. An organisational change will have implications for the technology of production, for the technology of communication, transport and so on. So it is worth the project manager's effort to establish the full technical reach of the project; this knowledge will be essential, anyway, in the identification of the expertise required by the project.

Social Component

What is usually not evident is the social component of the project, the component that refers to the people who will be affected by the project. The project manager needs to identify those so affected and consider their interests from their point of view.

A word of warning. The social component is very subtle; in fact, it can act in quite unexpected ways. A social logic operates which changes from situation to situation. That logic is often tied to the relationships between people, their shared values, their social conventions, their interests and so on. It is very difficult to deal with.

The social component involves culture, peer groups and self-image. It obviously applies to organisational change projects, to political campaigns, and so on. So there are projects where the social component is the most important part.

Projects that change production processes have a social impact. The introduction of new information technology that changes the way information is distributed within an organisation changes the degree of control and power people have over information. Someone who played a central role in the previous information distribution system may not be required in the new system; if they realise what is going to happen, they may well put up considerable opposition to the project.[37]

Sometimes the social organisation of labour feeds back to the technology and forces changes in it. A process may previously have required that one skill be brought in before another; however, this technical solution may be dropped in favour of another solution. An example from the building industry might be where blockwork, which is carried out by bricklayers, is dropped in favour of concrete walls, which can be constructed by the concretors who will also be laying the concrete floors. This solution means less moving and interchange of skilled personnel—just use concretors who will do the lot. Thus the social organisation of labour can affect the method of achieving the project.

An example of a project which suffered considerably because of the social component is the Redfern Mail Exchange, which was a semi-automated mail exchange. The project was part of a strategy adopted by the postal authorities in Sydney, Australia, to centralise mail sorting. It involved the introduction of electronic equipment which meant postal workers had to work in a different way. One interpretation of the failure is that the process moved mail-sorting from a batch mode (ie, a given amount of mail-sorting was to be done, with regular completion points) to a continuous mode (with essentially no end in sight). This led to so much industrial disruption at the mail exchange that the project was abandoned; it had worked at the technical level, that is, the machines worked, but failed at the social level.

One other subtle feature of the social component is the issue of moral or ethical values. The environmental movement clearly works in the area of values. A new product may cause unacceptable pollution, which leads to conflict on the project. But note the emergence of a new requirement in this area because of environmental value systems—the requirement that companies introducing new products also plan for their disposal. This is a significant extension of values in relation to the obligations of sellers, which will have considerable impact on the reach and extent of projects in the new product area. This development has emerged from the social component.

The project management technique called the Work Breakdown Structure (WBS) and its related methods of product and organisation breakdown structures are very useful in quickly and clearly identifying the extent and reach of a project. While not perfect, the WBS method is recommended in seeking to specify the full extent of the project. In setting up the WBS, the project manager should specifically identify the social component of the project. Another technique that helps highlight issues is the Fish Bone Diagram (or Ishikawa Chart) used in the quality management arena.[38]

6. Managing gaps

As the project clarification proceeds, it will become obvious that not all the expectations can be met, while unexpected consequences and problems will emerge. At the beginning of each phase, there will be both explicit and implicit expectations about its outcome. Gaps will emerge between what is expected and what is realistic, and compromises and changes will have to be made as the project proceeds. This needs careful management. This will be discussed in a later chapter under the topic of scope management but the sequence of managing the gaps starts here.

7. Estimating cost

Part of clarifying the project is deciding how much it will cost. It is important to recognise that the cost management process usually employed in project management is a satisfying strategy, not a cost-minimisation strategy. A cost target is set—the major cost targets being set now in the feasibility phase—and the designers try to come up with a solution that is within that cost target. A cost target or budget is set for the item being designed and the matter is progressed from there. Once the design has reached a point where the criteria are satisfied, there is little or no pressure on planners and designers to improve on it.

In most cases, this is satisfactory, but astute project managers, and certainly some hard-headed clients, do not accept this. Having achieved the goal, they move the goalposts and demand further savings. In other words, they set up another cost target or budget. One needs considerable authority to do this, to simply walk in and demand that design work continues in order to achieve a further reduction. One also needs to be confident that better designs can be achieved, as it can be quite expensive to send people off chasing savings if they cannot be realised. If the project manager does not feel comfortable with this dictatorial approach, a 'value management' process or review can be suggested. This review can then set up the new targets.[39]

The cost estimate produced in this phase should provide an overall cost target for the project and some limited distribution of those costs throughout the project. The number of items covered in the first budget will vary from project to project—on any substantial project, at least 15 items should be separated out and identified. On large projects where considerable project clarification work is done during the feasibility phase, the budget can cover hundreds of items.

There are various methods of arriving at a first cost estimate. The three main ones are cost comparison; setting an upper limit on the basis, say, of expected income; and building up a picture of the costs using very detailed explanations of how the project comes together.

The first method using a cost-comparison approach relies on comparing costs for the delivery of similar functions on other projects. This method is quite satisfactory in periods of low inflation, on projects of short duration and where the proposed project can be compared quite well with much of what has been done before. People who specialise in this area of estimating spend considerable effort in building up detailed understandings of the distribution of costs on existing projects so that they can apply these insights to the proposed project.

The second method takes a global approach: it attempts to set the overall cost for the project. That overall cost may be set by the amount of money available—so much money is available and therefore that is the upper limit on expenditure. This method usually identifies the expected income from the project and the required rate of return and, on this basis, a capital value is put on the project. Thus the overall budgeted cost figure is set.

The third method requires that one theoretically builds the project, that is, go through a theoretical exercise of putting the project together, and counting and costing all the resources used. This is a very demanding method which will probably require an extended feasibility phase.

A cost contingency should be included. It is important to recognise that the project is at a very early stage and difficulties are more than likely to be underestimated. It is quite likely that it is a psychological necessity that difficulties be underestimated (who would go into a project already feeling depressed?). The size of the contingency is of crucial importance here: contingencies somewhere between 20 and 50 per cent (and possibly even larger) need to be considered. If the cost estimate has been arrived at by building up cost components, then the contingency needs to be added; if it has been arrived at as a maximum figure, then the contingency needs to be taken away to give one a lower figure as the allocated cost.

The figure arrived at for the expected cost of the project should take on some significant, symbolic status. In the author's view, the expected final cost of the project in actual dollars spent (not in today's dollars or in final-year dollars) should become the openly stated and supported project cost. Significant commitment must be made to cost control—it just seems too difficult to try to stay committed to a moving dollar amount ('Yes, we will spend $10 million in actual dollars, but that is really less than $9.1 million in values of three years ago.'). There is one situation where this may not apply and that is where it might not be politically suitable (it may be very morally suitable but that is another matter) to reveal or predict the final dollar amount. On the other hand, it is quite fortuitous that the final cost of the Sydney Opera House was greatly underestimated—a project management failure leading to a great project success!

8. Estimating time

This is seen as the stuff of project management. This is where the networks are brought out and come into their own. While networks have been given a lot of attention from theorists, and some quite complicated and sophisticated models exist, one needs to approach them with some caution and scepticism.

At present, what is required is an overall estimate of the time for the project with some details about milestones. Again, depending on the level of detailed data available on the project, there is some choice as to the level of detail in the networks underpinning the time estimates.

As with cost estimates, there are various methods of arriving at the first time estimate. The three main ones are time comparison; setting a time on the basis of a desired or key milestone; and building up a picture of the durations using very detailed explanations of how the project comes together.

The first method relies on the availability of comparable date. The second method is based on the date of an expected market launch or some other date that is deemed important. There are many tax minimisation projects that have 30 June as their key date.

The third method is quite common; however, a knowledge of production rates is required and assumptions have to be made about the level of personnel available. In preparing these time networks, one will find that some features are built in that lead to averaging out. For instance, a production sequence may require two days' work, but time is needed afterwards to allow a chemical reaction to take place. Thus one is dealing with an item of longer duration than might be expected from the production rates if they were used alone.

There are projects where the project sequence beyond a particular point is almost completely unknown. In such cases, one may have to make a guess but it is important to be quite explicit about this.

Another feature of time is that it is quite strongly influenced by the level of effort. Some assumptions about production rates may lead to situations where there is not physically enough room for all to work on the project sequence; alternatively, there may not be enough skilled personnel available. Most of the computer programs available allow some smoothing of the peak demands for people.

One also needs a time contingency—again, as with cost, optimism is usually the order of the day and needs some control. Additionally, while it is rare that things are included that are not required, it is easy to forget something. A time contingency of up to 30 per cent is often appropriate.

The use of the contingency is actually a way of treating risk. While it is not very sophisticated, it has the merit of simplicity and gives a quick sense of the level of risk. It also has the advantage that people can relate better to 'Some time up our sleeves' than 'There is an 85 per cent probability that we will meet the time target'.

As with the cost estimate, one needs to build in psychological commitment to the time target. Thus the time target should be translated into the date of completion, not the period of time up to completion.

9. Documenting the proposed project

It is a very worthwhile discipline to build up the documentation describing the proposed project as the phase proceeds. On any project, particular questions will be raised and the answers and the reasons for them should be documented. An easy way to put this documentation in place is to produce a fortnightly or monthly report.

The project manager must anticipate and be aware of all sorts of pressures trying to change the definition of the project. It is very easy to change a decision made a few months ago if no record of that decision exists. Memories are not capable of holding every detail, and confusion can easily arise if no formal system of clarification exists. There are managers who, through not keeping track of the details, regularly cause confusion which they can only sort out because they hold a position of considerable power. Proper documentation will hold the decisions in place or at least force a formalised process to come into being if a change is to be introduced.

Many project managers hold the view that what has been decided at the beginning must be produced—this is quite incorrect. If, on the basis of later information, a better outcome is possible, then things should change. The job of the project manager is to head in a general direction and make alterations as required, but to make them in a controlled, careful, and fully considered manner. Proper documentation is required to manage this change process properly.

The trouble with documentation is that it takes time and effort. Documentation can reach a point where it seems to dominate the project manager's time; at critically busy times, it is tempting to put off record-keeping until 'later'. Thus it is necessary to develop quick, easy ways of documenting decisions: these do not have to come out of a laser printer in a businesslike font, they can be in handwriting on a photocopied sheet sent to the relevant people. Action sheets are an example and their use is recommended.

10. Finalising the proposal

The main output of the feasibility phase is the project proposal, which basically recommends or rejects the project. It essentially acts as the first baseline and may contain explicit briefs for the next phase. The feasibility team, under the direction of the project manager, produces the document, which provides or covers some of the following:

- a clear definition of client and project objectives
- a description of the project (may include drawings and other documents)
- a description of the acquisition sequence
- a description of the roles of various participants
- a report on the various tests applied, including demand studies
- the funding method
- time management
- cost management
- logistics management
- the contractual approach
- environmental (physical and social) strategy
- quality strategy
- safety strategy
- insurance strategy
- IR strategy
- information strategy
- sensitivity analysis and cost risks, etc.

Evaluation of the project proposal

The first difficulty here is that one will not be able to spend much time evaluating each of the potential projects; one simply does not have the resources, particularly time. Only fairly crude evaluations can be made in the effort to reduce the list of potential projects to the point where only a small number remain for closer examination. This process relies on good judgement, which will be greatly assisted by having a clear idea of the purpose of each project. When the field of potential projects has been narrowed down, certain issues need to be addressed and a number of specific tests applied in the further examination and evaluation of those projects.

Eliminating bias

There are at least two frameworks within which the project manager might be working. The first, and the more usual, is that the client wants the project (or a project close to what is being proposed) to happen, and the tests are to be treated as obstacles to be overcome. The other, and less frequent, situation is where there is a genuine search for an ideal or appropriate project and the tests are considered as serious guides or criteria.

The conditions under which this second situation might arise are difficult to imagine, unless the evaluation is being carried out for a third party, such as a public authority or a bank. The sequence required to get projects off the ground almost always needs the commitment of a project advocate. Commitment implies—and requires—bias. Bias is a necessary, but maybe not sufficient, condition for commitment to a project to exist. Recall that projects, despite all the show of technology, despite all the show of numbers, usually emerge within a political milieu—someone with an interest in the project has gathered sufficient agreement and support for it to get thus far. They are unlikely to simply lie down or go away—by this stage, the project means too much to them. So, with the exception of third parties, it is more than likely that the testing will take place within the first, that is, the biased, framework. This book does not advocate bias, but merely notes its existence and points out that special care is needed to handle it.

A word of warning: the project manager needs to keep a fairly clear picture in mind when the tests are applied; needs to be able to understand what they mean; and needs to know what they can be constructed to mean. But, foremost, the project manager must be on the look-out for danger signals. If the project simply does not stack up in terms of the client's interests, then the project manager must, and must be able to, advise the client accordingly. If it becomes clear, for instance, that there is no market for the project or that it will be impossible to get government approval, the project manager must forgo the advocacy role and return to a detached, professional, disinterested role. (It is worthwhile noting here that most professional associations would impose such an obligation on the project manager, as well as the obligation to take account of the community interest which may supersede the obligations to the client.)

Applying tests

The evaluation of the project proposal involves the application of tests. The tests provide data, and the project manager, client and others will have to assess the implications of that data for their own decision-making. Clearly, one's attitude to risk will significantly alter one's view of the results of the tests; for example, for some, the discounted cash flow (DCF) will support going ahead with the project, whereas for others it will indicate the opposite. The key point is that all that these tests do is provide data for decision-making; they do not replace decision-making. It would probably be more accurate to say that the tests only serve to confirm or deny decisions already taken.

Note that this methodology is similar to that of the design process described earlier—one makes a tentative choice and then examines the choice. The data from the tests may indicate ways in which the choice needs to be modified, but essentially one will obtain either confirmation or denial of a decision already taken.

The underlying theory to many of the tests is covered in the literature on general management theory which is very closely concerned with justifying investments, justifying expenditures on various ventures and so on. Thus much of the theoretical underpinning of the feasibility stage is well-documented outside the realm of project management, but it still applies to project management. Some of the theory that has arisen in project practice has been in the areas of the environment, sustainability, and the politics of protest.

For most projects, the following questions define the tests that might be applied. Specific projects may have questions that explore certain areas in much more detail and/or other questions. While the questions themselves may be simple, answering them is not.

1. Does the project satisfy the objectives?

The project has been proposed in order to satisfy some needs and wants. These should now be revisited and examined in terms of the project under consideration. It will usually be necessary to examine the strategic interests of those promoting the project and ensure that these interests are being served by the specific project now being examined.

It is actually very difficult to keep a project on track, to keep a project proposal closely tied to the needs and objectives of the client and other stakeholders. Gaining and maintaining an agreed definition of the objectives is a political and social skill requiring considerable listening and talking.

There is a clear need to decide degrees of quality. The project manager will find quite widely varying views as to what quality standards should be used in the project, and a great deal of money depends on those views. Differing opinions will emerge quite strongly in large, complex organisations where there are clear structural interests in how the project performs. Those who look after capital, or whose role in life is to reduce the expenditure of capital, will take a different view from those who wish to reduce maintenance and increase reliability. Production people will want different things from marketing people, and so on. These differences are important and will have a considerable influence on the cost of the project.

The project manager has to ensure that the objectives are compatible with the needs, and that the proposed project meets the objectives.

2. The story test

There are project stories, words that describe the project, its purpose and method of realisation; that describe who is involved and how they are involved; and that allocate and maintain psychological ownership. These stories are not written down—they exist in people's heads. They are, in fact, a set of understandings that exist between those involved in the project and develop as the project sequence proceeds. The project manager must ensure that the stories are compatible. (The use of the word 'story' in this context refers to a social construction of a supposed reality.)

While there are two ways of describing a project, the written one and the unwritten one, it is the latter that contains much of the vital social understanding that really keeps project participants together. The project manager must maintain agreement on the unwritten stories, because inconsistencies in people's understanding of it—of the project—will, at this stage, indicate a breakdown in the translation of the objectives to the proposed project.[40]

3. Will the project operate as expected?

This question might be a subset of the first question. However, it needs to be examined separately. It is concerned with the ongoing operation of the project after it is handed over to the users. The example given earlier of the Redfern Mail Exchange might have been seen as satisfying strategic objectives, if it could have been brought into successful operation. The purpose of this question is to force consideration of the viability of the operational phase of the project.

4. Can the proposed project be successfully brought into existence?

This question can be rephrased as 'Is it technically feasible?' It is about the physical or social possibility of the project coming into existence, not about the capacity of the project team to achieve it; that is the next question. (While projects may be technically possible, the project team may not able to bring it into existence.) Technical feasibility refers to two aspects: it refers to the project itself, and also to the method by which the project will be achieved.

Projects often have an element of technical uncertainty but go ahead nevertheless in the belief that problems will be solved as the project proceeds. The Manhattan Project is an example of a project that had this big question mark. However, even projects that appear simple, for example, product development projects, can have hidden risks in this area.

Remember that the project manager is not the expert. The viability of the technology and operational feasibility is usually within the area of engineering and science, but there is often risk associated with it. Since the problem for the project manager is to know which advice to rely on, procedures need to be established to get as much accurate information as possible. Essentially, it is recommended that the project manager assume that the technology does not work and demand proof to the contrary.

5. Does the project team have the requisite skills?
It is quite possible that, while the project is technically feasible, the assembled project team is incapable of achieving it. The necessary specialist skills might not be available, supplies may be insecure, or the required information systems may not be available. Assembling an appropriately skilled team is the responsibility of the project manager.

6. Is the project politically feasible?
Can the relevant government deliver effective approval? This aspect is probably more important in overseas projects, but it may be an issue at home if the project is a source of controversy. The placing of optical cables above-ground in Australian cities in the mid-nineties was a source of considerable disagreement, with one level of government in conflict with the exercise of power by another.

While this question must be asked in relation to projects that rely on some form of government approval, there is a whole range of politics outside the area formally recognised as political. It has to be asked and answered in relation to projects aimed at achieving organisational change: does the proposer of the change have the organisational clout to bring the change into existence? Will the promoters of the change be able to overcome the resistance?

7. Are the required physical resources available?
A fairly obvious question that is usually not very important, since most of the physical resources are available in the market. However, the existence of monopoly suppliers or supplies from politically sensitive areas should trigger more careful consideration.

Oddly enough, the answers to this question may have misled some people competing to design the Hong Kong and Shanghai Bank, to become at the time one of the world's most interesting buildings. Some designers examined the supply of building materials in Hong Kong and noted the absence of a steel industry but the existence of a concrete industry. This led some designers to offer a concrete building as a solution. However, a steel solution won with the supply of steel coming from overseas.

8. Do the benefits exceed the costs?
In practice, a cost-benefit analysis is produced to cover a wide range of justifications. As the name implies, a cost-benefit analysis tries to identify all the costs associated with a project and all the benefits. There is a need to draw boundaries around particular entities and do the cost-benefit analysis in relation to them alone. This would involve identifying their costs and their gains from the project. So many analyses can be done.

Usually only one is done which tries to include the costs to anyone and everyone, plus the environment; it also tries to identify all the benefits to anyone and everyone, plus the environment. Of course, some people carry more costs than others, and the same goes for benefits. The siting of an airport is an obvious example; it is no wonder that Sydney, for instance, has spent about 40 years examining alternative airport sites.

Great skill is needed to identify all the costs and benefits, and who will pay and who will gain. Basically, however, the outcome of the analysis is essentially a personal judgement or an opinion (admittedly an informed one). In other words, the cost-benefit analysis is a tool in the justification for the project.

Because so much opinion is involved, there is the opportunity to introduce bias into a cost-benefit analysis; in fact, it might be theoretically impossible to avoid such bias. Cost-benefit analyses cost money and time, and the resources available to it are limited. This means that judgements need to be made about what is important and about what aspect needs more investigation. There is then the presentation of the analysis, with all the possible ways of emphasising certain aspects of it that that entails. This is not to say that everyone who prepares such analyses is deliberately setting out to mislead or that they do in fact mislead. Bias is almost impossible to avoid.

The great service a cost-benefit analysis gives is that it forces many participants to take a wider view of the project and it highlights many problems that would crop up later. The cost-benefit analysis also helps identify the risks and helps people to decide whether or not a project is worthwhile. This does not mean that all the readers of the analysis will come to the same conclusion.[41]

Part of the cost-benefit analysis often includes a discounted cash flow (DCF). Often a good and quite accurate estimate can be obtained on the costs, but in the final analysis the expected income from the project is a guess—an educated guess, but nonetheless a guess.

This emphasis on the difficulties in establishing the income of a project is made because one will often be confronted with a very elaborate DCF analysis showing the derivation of costs in great detail with a single line showing income. This is the most vulnerable line, the line needing lots of analysis. When examining the DCF, look carefully at the predicted income stream.

DCF techniques are very easy to manipulate in order to produce the desired answer. The author has seen proposals for the introduction of Computer Aided Drafting (CAD) systems showing internal rates of return of up to 4000 per cent. A quick look at the analysis showed that a slight change in the income assumptions threw the cash flow into a large loss!

Sensitivity analyses need to be carried out, with variations in the assumptions tested. It is recommended that quite large variations be examined; for example, interest rates doubled almost overnight in the early 1970s, actual income streams can vary enormously, and so on. If analyses indicate danger, contingency plans are necessary. Contingency plans must look to the possibility that other projects will experience the same problems at the same time, thus leading to the possibility of quite unsatisfactory market conditions.

Underpinning the income data is a view on likely future demand: this falls into the area of marketing. Many readers will be familiar with marketing and its concepts. Remember that the likely future demand for the project is a crucial central question that cannot be exactly answered but must be carefully considered. Many of the remaining tests are irrelevant if the demand is not there.

9. Can the project sequence be funded?
This is a totally different question from the last one. While some quite profitable projects never get off the ground because they can't get funding, other quite unprofitable projects have been funded.

The state of the project organisation, its access to resources, and its plans for the allocation of its resources all have to inform the feasibility stage. If the project is a company project, then the company needs to examine its cash flow and assure itself that it can supply the funds when required. If it cannot fund the project directly, then it will have to turn to one or more financial institutions.

The project may have to be funded by special legal entities in order to avoid certain taxation issues or to make it clear how the funding is provided. The funding of some major infrastructure projects has introduced some interesting legal vehicles, not to mention secrecy.

The funding needs to be flexible to cope with different rates of progress on the project.

10. Is the project proposal and sequence legal?

Leaving aside the obviously criminal projects, which will proceed despite the law, the legal issues facing a project may need examination. In some project areas, there are well-known legal issues. For example, in the pharmaceutical industry, there are defined stages and processes for the introduction of drugs; in the construction development area, there are also defined legal stages to be followed in gaining approval for a development.

11. Is the project environmentally sound?

This is a huge area which has expanded dramatically in the last 20 years. It covers a wide area including direct pollution, the use of energy and the reuse of resources. The word 'pollution' has taken on a wide range of meanings—it can mean chemical pollution as well as visual and sound pollution.

The reuse of materials and the minimisation of the use of energy are issues that are now deeply embedded in the minds of much of the industrialised world. As mentioned earlier, these values are extending the perceived legal liabilities of those introducing new products.

A major effort to answer this question comes under the guise of an Environmental Impact Statement (EIS). However, many EISs are produced at a later stage in the project, often in the planning and design phase.

12. Is the project acceptable to the community?

This goes beyond environmental issues; it reaches into the values people have and their views of what goes on in their community. It ranges from simple self-interest to the imposition of moral values. Many community projects, such as a youth drop-in centre, may be resisted by local residents who see their enjoyment of their suburb threatened. The whole issue of censorship on the Internet is an interesting conflict between values and self-interest.

13. Is the risk level acceptable?

There is always risk—the question is whether or not the level of risk is acceptable to the project participants and to the community.[42]

Risk management is increasingly discussed in project management. Note that while taking risk into account can reduce it, it cannot eliminate it entirely. The process of risk management involves identifying the risk, assessing it, developing ways to reduce it, and then deciding how it will be distributed. Insurance is an example of giving the risk to an insurance company, or of displacing the risk. An important principle that appears to be developing is that risk distribution should be done to a significant extent on the basis of capacity to bear the risk.

It must of course be noted that not all the risks may be identified, so that in fact there is always a risk that cannot be transferred.

At this point in the project's development, there is a need to decide whether or not the risks are acceptable to the client. If the client sees the risks as too large, a risk management process may be able to identify how the risk can be distributed in a way that is acceptable. Otherwise it may be necessary to abandon the project.

In business, there are some owners who will 'bet the company' on a project; a strategy of regularly betting the company is probably doomed to failure.

On projects, one finds there is financial risk (which can be grossly magnified by foreign currency borrowings without hedging), there will be a time risk, there will be a technology risk and there will be legal risk, among many others.

14. Can the project be delivered on time?
On some projects, this is not critical, but for others absolutely crucial. The test here is about the likelihood of the project being achieved within the desired timeframe. Sophisticated network techniques have been developed to address this question; details of these are available in the specialist literature. A useful idea that should be kept in mind when considering the time issue, whether or not one is using a sophisticated technique, is to not simply look for the critical path but to also look for those activities that might easily become part of a critical path.[43]

Using the test results
The first use of the tests is to help in the selection of the appropriate project from a range of project proposals; the second use is to act as a vital safeguard against the selection of an inappropriate project. Some of the tests increase the chances of success, such as those concerned with objectives, while others reduce the risk of failure, such as those that test demand for the project. These tests do not guarantee success but they probably significantly reduce the chances of catastrophic failure. They allow rationality to be applied to the issue of choice but, as said before, they do not make the choice.

These tests can be applied at any stage in the project life cycle. Here, they are being applied very early, during the feasibility phase. Applying them again at the end of the planning and design phase is a useful and desirable check. Underlying this testing process is a gradual stepping along the project sequence, making further commitments as the checks confirm that all is satisfactory. In fact, the project life cycle approach is designed to allow this tentative movement along the project sequence.

Some of the tests can be entirely dealt with in the feasibility phase, such as examinations of cost/benefit ratios, while others have to be dealt with in later phases. For example many development projects can only complete their environmental and political tests in the planning and design phase.

Acquisition of the project

Now that it is known what is wanted, and that what is wanted is a viable, appropriate project, the next question is how to bring it about. Acquiring the project means:

- **deciding the project sequence**
- **deciding the project phases**
- **deciding the infrastructure required**
- **deciding the project organisation**
- **deciding the contractual relationships**
- **managing the overview of the project sequence**

To some extent 'deciding the infrastructure required' and 'deciding the project organisation' overlap, but they are separated here for the purpose of clarity. All of the above interact as the work proceeds but are separated for purposes of explanation.

The project sequence

This part focuses on the sequence of activities required to achieve the project. On a software project, one might have to decide on functionality, then decide on detail outputs required, then decide on data sources, do various other things and then start coding. This sequence of steps is independent of staffing, financing and various other things—it represents what is technically required to achieve the project. In an organisational change project, sequence steps might include deciding the messages to be conveyed to staff in training, prepare staff training modules, organise staff availability, deliver courses, check the status of messages received by staff, and announce the details of the re-organisation. Even for so-called 'soft projects', there will be a sequence of activities that are technically necessary.

One particular sequence needs to be identified for use in acquiring the project. The sequence does not have to be unique to each project, but only one sequence can be chosen. In fact, it will rarely be unique—many different sequences can be found but some are more effective than others.

As mentioned earlier when discussing the project life cycle, it is not now necessary to identify all the activities required. For many projects, this would be impossible and grossly inefficient. One needs to identify the necessary sequence of activities sufficient to allow one to proceed. If it is appropriate, then the identification of the sequence can stay at the level of phases, rather than activities within the phases. For some projects, it may only be possible to identify a milestone which may, at a minimum, be simply a point at which a certain amount of money or time runs out.

The key advantage to identifying the technically necessary activities is that the path to acquiring the project is pushed to the forefront of the planning process and one then decides other items with a view to facilitating progress along this path. All the other elements such as financing and other aspects of infrastructure hang off this sequence. Aspects of infrastructure are there to promote and facilitate this sequence, not to dictate it. Unfortunately, in practice this will not be that easy—the infrastructure will influence the choice of the technically necessary sequence and vice versa. Despite this, the focus must be kept on moving along the sequence.

A design or problem-solving process, as described earlier, is used to identify the technically necessary sequence; and there are two ways of examining and presenting the work associated with it. The first is the Work Breakdown Structure (WBS), and the second is any one of a range of network methods. At this early stage, a simple precedence diagram is probably appropriate.

While the WBS may not be the best way, it is useful in that it gives a 'whole of project' outlook. The WBS will not be completed in this part of acquiring the project; later steps will further add to it. This is very important to keep in mind because we require that the WBS illustrate the full extent of the project, including the setting up and running of the infrastructure.

Deciding on the overall approach

What must be sorted out now is the overall approach. The first issue concerns what is commonly called the 'make or buy' decision, and the second concerns whether or not the planning is to be separate from or form part of the implementation of the project. This leads directly to the timing of the main tender.

Make or buy

Two decisions are involved here. The first decision concerns what parts of the project can be bought 'off the shelf' as against having them specially made. Can the client simply buy something that satisfies the needs? While it may not be possible to buy the whole project off the shelf, it will almost certainly be possible to buy part of it. The United States defence establishment has been ridiculed for paying huge prices for a one-off item, while an equivalent, mass-produced in the marketplace, was available at a fraction of the cost.

The 'make or buy' decision must weigh up the advantages and disadvantages of each option. To make the cost comparison, one often uses discounted cash flow techniques. But not to be forgotten are other issues such as function, delivery time and reliability. In general, the decision to make should not be a marginal one, The 'make' option should not be chosen unless there is a significant advantage in doing so; it should clearly be the best decision that almost demands to be followed.

The second decision concerns the extent to which one should use the services and products of outside organisations. Most organisations do go outside, ie, buy outside services and supplies, to acquire their project or a significant part of it. Even some quite large organisations who think of themselves as doing the project themselves, organisations such as utilities, will in fact acquire much of the project from outside.

Even if an organisation can do some of the work required, it still may not be appropriate for them to do it; some organisations can make better use of their resources doing things other than the project.

However, all organisations should carefully weigh up to what extent they want the function of project management to go outside. There certainly is a temptation to hand it all over and take the turnkey approach, that is, let the supplier do it all so that the client only has to accept the project and turn it on with the key. In areas where the project is to become a core part of the operational activity of the organisation, there is good reason to keep the project management within the client organisation so that it can maintain control over its core technology. An organisation does not need to be on top of all the technology it uses, but it needs to be on top of its core technology.

A word of warning. On a one-off basis, acquiring the job from outside may seem more expensive, particularly if the client organisation sees itself as having some skills in the area. The organisation may be tempted to save the money that would otherwise go to the outside supplier as its profit by doing the work itself. But remember, it is a one-off activity, and afterwards the skills and resources specially acquired may be of little use—in fact, they may become a liability.

Design and supply

This goes under various titles, one of which is 'design and construct'. This approach is very close to the turnkey approach (the turnkey approach would further extend to the handover or commissioning phase).

The decision here is whether or not to separate out design from supply. If one separates them, the client carries the main responsibility for the management of the planning and design phase. The alternative is to leave the supplier with both design and the supply. In this case, the supplier is mainly responsible for the design phase.

Several things favour entering into a design and supply (or a design and construct) arrangement. The technology may be too complex and the supplier has particular expertise; the client organisation is unable to supply the co-ordination and project management functions; the supplier can (or is expected to) bring the advantages of co-ordination of design and supply; it is industry practice. Where none of these apply, however, there is a tendency to separate planning and design from supply and construction.

With the exception of quite comprehensive turnkey supply projects, one will find on most projects that the procurement system consists of a mixture—some components will be obtained by design and supply while others will be obtained by design and then supply.

Novation

A novation contract is a new form of contract: it starts out with planning and design separate from supply and construction, and later brings them together. The design process proceeds, under the direct control of the client, to a particular point at which the design consultants are transferred to the supplier or contracting organisation. The design consultants then treat the supplier or contractor as a client, with the supplier and contractor dealing directly with the project's client.

Going to tender

The main reason one needs to think about tenders at this stage is because the form of the main tender has a profound effect on how responsibility for the project is distributed and managed. The main tender is the one which incorporates the work for the implementation (and pre-implementation) phase. If planning and design are kept together with supply and construction, then one will have to go to tender immediately after this feasibility phase, or very soon thereafter. That tender will cover the planning and design, pre-implementation and implementation phases. If planning and design are separate from supply and construction, then one enters a set of arrangements to carry out the design and, following that, one goes to tender for the supply and construction.

One can go to tender for the planning and design work, which will not be as big a tender as the one covering implementation.

There is another point at which one could go to tender and that is between the concept and documentation stages of the planning and design phase (see Chapter 5). In this case, one keeps the conceptual work under close client scrutiny and then hands over the functions that are essentially production to the main contractor or supplier.

There are always several tenders on projects: the main tender may cover 70 per cent of the work, while some of the smaller tenders for minor services may cover less than one per cent of the work. The acquisition sequence involving tenders is often as follows:

- identify potential suppliers
- decide type of contract
- call for bids
- assess the bids
- negotiate details
- award the contract

From the supplier's point of view, it is necessary to ensure that the entity calling the bids has the wherewithal to pay the costs associated with the work. This is a growing issue, particularly when bids are made across national boundaries.

The key issues here are deciding the type of contract and assessing the bids.

Potential suppliers

For people in a particular industry, identifying the potential suppliers is very easy. For those outside the industry and where projects are so new that there is no history of suppliers, it can be difficult. If one is in completely new territory, then a short-list must be drawn up of potential suppliers with whom negotiations can begin.

A short-list is also usually drawn up for large, complex or risky projects so as to exclude those who, through ignorance, may come in with the lowest price and cause havoc later in the project when their incompetence is fully exposed. Legally, they may be carrying the risk but in practice the risk will be carried by the client.

Setting up the short-list can be done by public invitation; by inviting companies and organisations to express interest in making a bid to do the work. Certain criteria will be applied in order to make a selection from those companies or organisations. Those selected, the short-list (in some cases called a panel), are invited to make bids to do the work.

In these procedures, there is a growing trend to transparent processes, most particularly in the public sector. From the project manager's point of view, this will mean that reasons can be produced to justify a choice and these reasons will have to be acceptable within the framework of open competition and fairness.

In practice, there are many instances of procedures that are essentially illegal, improper or immoral. One example is the widespread practice of telling preferred subcontractors that the job is theirs if they meet the best price obtained in open tender. Other subcontractors are asked to submit bids and carry the expense of tendering, simply to serve as a pricing mechanism for a party who is not involved in the tendering process, and who does not have to meet the costs of bidding. It is surprising how many people think this is reasonable practice.

Type of contract

This will be dealt with below under contractual matters. Essentially, it is a choice on the spectrum from fixed price to cost reimbursable.

Call for bids

This requires announcing the bid in such a way that those whose bids are sought will be notified. Each industry has its own way of doing this. For instance, *The Sydney Morning Herald* in Sydney has local government advertisements calling for bids every Tuesday.

In some industries, electronic methods are used, where whoever is calling the bids sends out its invitation to bid via computer to the various potential suppliers. They in turn respond via computer and the whole process can move quite quickly—in a matter of days (possibly hours) as against weeks. The technology underpinning this is Electronic Data Interchange or EDI, sometimes called electronic commerce. This method appears quite effective where the entity calling the bids is large and powerful and able to impose its will to some extent. There is considerable work going on in this area; in time, one will find that there are open standards capable of supporting quite open bidding processes.

While the above seems simple enough, it can be complicated by changes that need to be introduced into the scope of work or into some of the conditions. In setting up the process, it is worthwhile having an explicit change process incorporated into the documents calling for bids. These could, for example, specify how changes, if they occur, will be advised.

One will also need to have a system for handling inquiries about the tender. If a contact name is given, answers can be consistent. If one careful bidder raises genuine problems, as a result of their more insightful analysis of invitation documents, there is the dilemma as to whether or not other tenderers should be informed. One should make one's views on this explicit at the time of calling tenders.

On large or complex projects, the client may announce a meeting to discuss the forthcoming tender with possible bidders. An example of this was where the Sydney Water Board, in the late 1980s, called a meeting of interested parties to answer queries and provide more detailed advice on a forthcoming tender to outsource an entire function.

Such meetings allow the client to provide some more contextual information and deal with preconceptions that may exist.

Assessment of the bids

This is quite a complex matter. Ideally, the tender should be in a form that makes it as easy as possible to compare like with like. From the bidder's point of view, however, it is a selling exercise, and care is taken to submit a tender in a format that is acceptable (ie, that conforms to the stated requirements) but which is in some way quite different from all the others. In other words, the selling objective is to obfuscate direct cost comparison, unless of course cost is a clear winner for the bidder.

No matter how hard one tries to avoid it, there will be differences to deal with. The prices will be distributed differently in different offers. Even when the total amounts are very close, and it is surprising how close the lowest three offers often are, one will find quite significant differences in timing of payment and in dealing with variations to the contract. Many assessors set up tables using weighted factors to allow them to construct an index to allow some comparison. These only go so far in real terms but they are quite useful in defending a decision or choice.

In the public sector, it is quite an onerous task subject to query and appeal. Where a subjective judgement is to be made, it is often important to set up a panel, so that the procedure for selecting the panel will be the main defence against queries or suggestions of bias. Transparency as to procedure becomes an important feature in the public selection process. Evaluation should be on the basis of the technical competence of the offeree, and on the reasonableness of the time and cost components of their offer. Registered quality assurance systems are also a consideration.

The offers will contain differences in the way risk is treated, a very difficult area to assess. Some offers will be ignorant of the risks and will in fact not have included the full costs of them. In assessing the distribution of risk, the assessor, as pointed out before, will need to be mindful of the capacity of the offeree to take on the risk.

Negotiation of details

Following evaluation of the offers a successful bidder is chosen. It is then quite common for further negotiations to take place, essentially to deal with minor matters. When these are completed, the successful tenderer signs a contract.

When one is a bidder and tenders have closed, the silence from the party assessing the tender is quite disheartening. If one is close or likely to get the contract, there is an invitation to meet to negotiate or clarify some matters. Negotiations and clarification always seem necessary. Being called in to clarify is a good sign—not being called in and hearing that the competition has been called in is not!

Among the items to be negotiated will be changes that have occurred since tenders closed, the exact form of some of the contractual documentation, details of various guarantees, clarification of some conflicting data and agreement on how some issues are to be interpreted.

Award of the contract

The contract is signed—in most cases, the project manager will sign as agent for the client; otherwise, the client signs on its own behalf.

Addressing other sequence issues

There are particular issues whose importance will vary from project to project; some of them need to be addressed at this stage.

Environmental issue strategy

This is not a surprise. If the project will have an impact on the physical external environment, then the method by which various issues can be addressed will have to be developed. Clearly, one must identify the issues and the skills needed to deal with them, identify who is involved and their interests and positions, and identify the steps to be taken to solve the actual and potential problems.

This area is broadening in influence and power. The definition of what is acceptable is changing and probably extending, so the project manager must be able to anticipate issues as well as identify those already established and incorporate their management into the project sequence.

Quality policy and program

There are two quite separate but interacting issues here. First what is required must be identified and then how it can be reliably delivered.

The issue of what is in the project has already been discussed; what needs to be thought about here is how the quality can be assured. For some projects, the statistical methods can be applied but only in a limited way. Because one is dealing with the one-off situation, one has to go beyond statistics. The Total Quality Management (TQM) concepts such as satisfying the downstream customer are very important in project management and should be recognised while developing the project sequence.[44]

ISO 9000 has moved into the area of project management. While it is probably still too early to pass judgement, some concern has been expressed at the way a paperchase has developed without any noticeable effect on quality. The costs associated with quite inappropriate applications of quality management are now being identified. There is no doubt that quality management in some form is here to stay but it is more subtle than many recognise.[45]

Safety policy

The safety issues to be considered here are those involved in project acquisition. For most projects, the safety issues of acquisition can be dealt with in the next and subsequent phases. There are some projects, however, where safety is a serious issue requiring immediate consideration. Projects involving nuclear products clearly have safety implications that need to be addressed very early.

The project sequence should allow for the safety checking of the product itself. For example, in the production of toys, there should be some steps in the sequence that are concerned with checking the safety of the product.

Legal strategy

In recent years, many law firms have recognised that projects constitute a large area of business and have begun to pay attention to it. Some of the legal and quasi-legal issues that should be noted in the feasibility phase are ways of dealing with regulatory bodies (deciding the form and timing of submissions), ways of approaching negotiations, ways of offering product to the public (eg, licences for software, as against ownership), and so on. To satisfy these requirements means identifying the appropriate project sequence steps.

Logistics strategy

Logistics is about getting things in place or having them available as required. This is clearly important in the setting-up of the project but is also of vital importance when the ongoing maintenance and operation of the project has to be supported by the client after acquisition or by the manufacturers who offer spare parts and after-sales service. On some projects, this requires a lot of work, perhaps more than the acquisition of the project itself.

Industrial relations strategy

Industrial relations covers identification and management of the terms and conditions of employment. The need for this can turn up in a whole range of areas; it obviously needs consideration in organisational change projects. An industrial relations strategy will certainly be required on projects where there is pressure for completion (such as getting an Olympic Games facility ready on time), on projects where some groups are going to control some vital resource (CAD operators) and on projects where different groups are going to be brought together.

Design development strategy

This is often a central issue subsumed in other activities and is often not given explicit recognition. By design development strategy is meant how one acquires particular skills or how one actually approaches the design task, or both. In the fashion industry, in the film industry, and in all the industries where design comes together with technology, people are known for their design abilities. How one uses them to do the design is part of the design development strategy; the necessary steps can be fairly easily incorporated into the task network.

In acquiring the particular skills to design a product or project, one might simply decide to go for the big-name designer and recommend that their involvement be negotiated. It is noteworthy that Renzo Piano, the world-famous architect, has been selected to do a major building in Sydney. This commission is not just going to any architect but to one of the stars of the profession. On the other hand, one might organise a competition and get an Utzon.

One can go overseas and import the design skills, and thus gain input from other cultures. A Japanese car company is reputed to have decided to acquire some of its designs by hiring or commissioning Italian design companies—a way of acquiring a different cultural history and approach to design.

The more usual situation is that one has to select one of the local designers, of which there is usually quite a number.

Approaching the design task is probably a little more difficult to identify and isolate. One issue might be the attitude to the length of time allocated to design in the project time-scale. Another might be the attitude and approach to designers—are they to behave 'normally' or are they to be allowed a certain latitude of behaviour. Advertising agencies are interesting organisations in this regard.

The attitude to adopt will not turn up in the project sequence but in the intangible part of the ongoing management; it becomes part of the project infrastructure.

Equipment and systems

Projects usually require equipment; some may need special equipment. The acquisition of specialised equipment for the production of a project has the attendant risk of there being no use for it after project completion. This is a major issue for companies when they consider their approach to the technical requirements of acquiring a project. Thus this is both a project and a company issue.

Acquiring the right equipment can be a significant advantage in getting projects in a particular area. In the engineering construction industry, when an Australian construction company foresaw growth in the demand for road infrastructure, it acquired very specialised and efficient equipment. It was then able to successfully outbid much of its competition for work that subsequently came along. It had both a time advantage (it already had the equipment) and a cost advantage.

In this phase, one should decide if there is any special equipment which could be used as part of the project or which could be incorporated into it. If it is decided to use this equipment, it is essential to make sure that the appropriate lead times can be handled. In deciding to set up a production line, one might evaluate the use of NC (numerically controlled) machinery as part of that line. This may involve the identification of a significant number of steps in the project sequence. Lead times need to be recognised very early on, as early as possible, and actioned. There is a clash here with design requirements. One of the problems for the project manager is that if the specialised equipment is outside his or her experience, the need for it may not be flagged by the specialists; they simply assume that appropriate arrangements are being made.

Wise thinking about the scheduling of equipment occurred with the new Fairfax Printing Press at Chullora, Sydney. The project team installed the major presses very early on in the project, rather than wait until all the surrounding building work was complete; this foresight probably saved close to a year in the production of the facility.

Prototyping

Prototyping work must be scheduled. The thing to recognise about prototyping is that it is essentially a learning experience; the extent of repeats in the form of refinements can be difficult to predict.

The project phases

The main purpose of phases is to break down the project sequence into manageable parts and to simplify the management task by grouping like work together. The project manager needs to decide the boundaries of the phases and, most importantly, the deliverables from each phase. It is probably best to define a boundary to a phase on the basis of a clearly defined deliverable. Each phase can then be allocated a time and cost budget.

The project infrastructure

The project has tasks to be completed which are allocated to phases; the phases have to be supported by managerial activity which, in turn, has to be supported by the project infrastructure.

The project infrastructure is necessary for the project acquisition but it does not form part of the final project—much of it is necessary for the tasks to be carried out. The main parts of the infrastructure are finance for the project sequence, insurance, the project information system, communication systems, the project office and the usual services associated with an office, such as the availability of skilled and unskilled human resources, equipment and systems, the manufacturing systems for the materials required by the project, transport systems and so on. It is a vast range of support available to projects, much of it unrecognised.

The infrastructure also has intangible parts, such as political support for the project, a willingness for people to participate and the legal system. There are things that may or may not form part of the infrastructure but which are not part of the project, for example, management and problem-solving methodologies. The project manager must ensure that they are recognised and accounted for in the feasibility study.

The project manager needs to identify the critical project infrastructure and evaluate whether or not it exists and whether or not it will be available.

The project organisation

There are two main control paths to note in a project organisation: there is the formal, legal substantive path of control and there is the practical path set up to manage the situation. Both of these paths may have legal force or only one may have legal force.

Despite all the talk that goes on about lack of hierarchy on projects there is, or should be, a clear organisation. Various teams will be working on parts of the project. The work of these teams will be co-ordinated by project meetings, by active interface management by the project manager, and by the project management team.

Establishing the organisational framework

There are four main structures within the organisational framework. In the first, most of the work is done within one department as part of its ongoing work. The second involves the establishment of a task force; the third, a project organisation; and the fourth, a matrix structure.

The task force applies to an organisation which needs to manage a one-off project every now and then; under a project manager, the task force carries out the project management function. A project organisation is where all the staff are attached to one project or another; the organisation consists of a set of project teams each running its own project. This model is fairly common in design consultants' offices where the staff are allocated to a job for as long as the job is there; it is not very common outside the sphere of specialist consultants. The matrix structure attempts to combine functional hierarchy with a project organisation. It is employed by fairly large companies which run projects but need to maintain their functional expertise.

What is common to the last three structures is that each is a different device used in the assembly of the project team. The task force is a direct set-up of a special team, the project organisation is a flexible way of organising project teams using a set group of employees, and the matrix is a way of providing a team and maintaining a functional hierarchy.

The teams thus assembled can be either the project team controlling the overall project or they can be one of the autonomous teams on the project.

Working with autonomous teams

Projects require that various islands of autonomous action are set up and that these islands are co-ordinated so that their work comes together properly. The key to getting a lot of work started is to get as many autonomous teams as possible working on the project as soon as possible. It must be assumed that the teams will manage their own internal work, and that the project manager will be responsible for their full and proper interlinking. On larger projects, this task of interface management is too much for one person and is carried out by a team reporting directly to the project manager.

It is important to understand the assumption that the teams will be autonomous, that is, they will manage their own internal work. This is not an assumption that would be left unchallenged in a stable organisational form, where it would be part of the management's function to improve internal teamwork. With projects, however, the project manager often must assume the competence of the teams, or build ways to gain assurance.

Many of the teams will exist as a result of contracts. There will be various design teams, various trade skilled subcontractor teams, and so on, who exist on the basis of a contractual relationship. Their competence is assumed—in many cases, they will be the experts. Thus, given the assumption that these teams possess both competence and expertise, it is difficult for the project manager to directly intervene in their internal affairs. Not only is the expertise a barrier but so also is the contractual relationship.

Access to the internal working of teams

The project management team must have clear agreement on the output that can be expected from the autonomous teams. It needs to monitor that output quite closely in terms of quality and time and, to some extent, cost (legally, the cost is the autonomous team's business but any failure on its part will leave the project with a cost, thus ending up in the client's lap). On some jobs, the agreed team outputs will not be deemed sufficient to reduce risk and later, when dealing with design teams and key subcontractors, it will be recommended that the contracts grant the project team access to the internal processes of the autonomous teams. This internal access is to allow closer monitoring and to assist in interface management, but certainly not to take over the work.

Information needs lead to teams

It may seem odd that teams seem to be the dominant theme of the project organisation. There is at least one good reason. On projects, large amounts of data need to be assembled and processed quickly, a large and complex task requiring lots of effort. There is a romantic notion that everyone should have access to all this data at all times; that in some way it should be possible to deal with any part of it at any time in a totally integrated way. This is practically impossible. What is practically necessary is that a lot of data gets processed locally, in one area, by one team, and then the results of this processing are transferred in blocks to other locations. Teams fit very easily into this method of managing data—in fact, they are very good at operating in this way.

Social needs of teams

Teams also have the advantage that they provide a response to social needs. While there are good and bad teams, there is little dispute about their social function. So, for some fundamental human reasons, there are good reasons to expect teams to perform well.

Choosing the project organisation

The author recommends that those charged with establishing the project organisation look at the information demands and the physical work-load that will arise in each of the phases. In the early phases, there is the need to process a lot of data. This leads to a set of teams clustered around the project team itself. This cluster takes the form of a very flat hierarchy where each element in the hierarchy is a team. The Likert model seems appropriate here.[46]

As one moves through the phases, one sees an increase in the execution of work as against the planning of work. Detail planning will still be there but the major planning efforts are complete. Now one finds that a deeper hierarchical structure develops that is a mixture of teams and functional hierarchy. In many cases, the work in the later phases will be repetitive and therefore more amenable to the development of a hierarchical rather than a flatter team approach.

No matter how the teams are set up and no matter how they cluster, one still needs a formal hierarchical path for the confirmation of decisions. In effect, one is dealing with different models operating together.

The authority hierarchy runs in parallel and, in some ways, in contradiction to the team model which will have its formal and informal social structure. But the message is that the information-processing needs dominate, and must have a clear line of authority.

The contractual relationships

The legal relationship between participants clearly has an impact on the form of project organisation that should be chosen. A contract for design and supply leads to quite different organisational structures from a contract for design and a separate contract for supply. In fact, much of the organisational structure is expressed by means of the contractual relationships.

Now turning more specifically to the contractual relationships, there are essentially two main types: fixed-price or lump-sum contracts and cost-reimbursable contracts. The fixed-price or lump-sum contract is closely related to a unit-price contract. In a unit-price contract, the nature of the risk in relation to quantity is modified. The cost-reimbursable contract sometimes takes the form of cost plus fee (which may be a percentage or a fixed amount), or even cost plus incentive.

Some argue that there is no such thing as a fixed-price or lump-sum contract. The basis of this argument is that changes will always be introduced and hence the price varies. The extent to which the price will vary is often related to the state of demand in the economy. During downturns, lump-sum contracts survive better than during boom periods.

If it is possible to fully define the work, one should move in the direction of fixed-price contracts. If much of the work has yet to be defined, one should move towards a cost-plus contract. In the initial phases, it would be reasonable to expect to see cost-plus contracts, and more and more fixed-price contracts emerging in the later phases.

The fundamental issue in relation to lump-sum versus cost-plus is the distribution of risk. Competent suppliers would be expected to increase their prices to match increased risks; incompetent suppliers may not even recognise the risk, a matter of concern. Some contractual relationships have succeeded in combining risk and incentive, such as guaranteed maximum price with a share of the savings below this price.

The client usually issues the contracts and the suppliers accept those contracts. The client needs to carefully consider how many contracts it is able to handle; a contract is not a passive item, it requires ongoing management. The purpose of using main contractors is to reduce the number of contracts the client has to manage and to concentrate the responsibility and risk.

Most industries would have standard forms of contract. In choosing a contract, the project manager should look at the industry standards and seek advice. Each of the standard contracts within an industry have various biases and peculiarities.

Since specifications form part of the contract, they need to be determined. The main choice in specifications is between performance specifications on the one hand and 'detailed' specifications on the other. In a performance specification, requirements are specifically stated and it is up to the supplier to decide how to do the job. It would be reasonable to define a market research specification in performance terms rather than tell the market researchers how to do their job. With performance specifications, it is hoped that the expertise of the suppliers will lead to savings. 'Detailed' specifications say exactly what has to be done and *how* it has to be done; the advantage here is that it is easier to know when the job is done, and control is easier.

Contracts in project management, as in general management, take the usual form of offer, acceptance, consideration, etc. The documents forming or evidencing the contract will vary from project to project. Typically, there is a general form of contract (fairly typical for that industry dealing with the usual range of project and contractual issues, such as names of parties, conditions of contracts, how variations are to be handled, etc) and a specification which will describe the specifics of the work. The specification can take one of a range of forms and may or may not include drawings.

One of the contractual forms that should be understood and may need to be used is the form of contract applicable to agency. The nature and extent of this form will probably have an important influence on how the practice of project management develops. This is particularly the situation where a professional project manager manages the project sequence on behalf of a client (as against an employee acting for a client).

The client may contract directly with suppliers and actively manage the contracts. In this case, an employee of the client acts as the project manager and the relationship between the client and the project manager is that of employer and employee.

An alternative is that the client appoints a project manager to act as its agent and to enter contracts on its behalf. While the contracts are actually between the suppliers and the client, for all practical purposes, the project manager stands in the place of the client. Here there are considerable elements of the professional duty of care.

Another form of contract is where a client enters into a contract with the project manager for the acquisition of the project; this is the main contract. The project manager then contracts various parts of the work to subcontractors; on each of these contracts, the principal is stated to be the project manager. In this case, the contract between the client and the project manager is more in the form of a contract between a buyer and a seller. There may be one difference here, and that relates to the extent to which the relationship is expected to contain elements of the professional duty of care.

The precise extent of the duty and liability of the project manager is a grey area. Many contracts for project management services contain significant disclaimers, their basis being that the effectiveness of the project manager is directly dependent on the performance of the client. To what extent this disclaimer works will become clearer with time.

Management of the project sequence

Management of the project sequence must look after a number of quite distinct but interacting areas:

- the management of the ongoing work of a phase (say, a feasibility team managing its work of the feasibility phase);
- setting things up so that there can be management of the ongoing work of a phase (say, the client appoints the feasibility team);
- the maintenance of an overview of the whole project sequence (it is quite likely that the feasibility team needs to provide a plan for all of this); and
- the work of ensuring that the project sequence is appropriately placed in the environment (for example, managing threats to the project).

In the feasibility phase, there are two levels of management. The upper level is the client, and the other, the feasibility team. The feasibility team would be reasonably expected to cover all the above areas except for the second, setting things up. The client would set up the team and expect the team to look after the rest in so far as it operates in the feasibility phase. However, we will describe the work of the first two areas—the management of the ongoing work of a phase and setting things up—in the next section on staff appointments.

In the feasibility phase, much of the work associated with the last two areas— maintaining an overview of the project sequence and ensuring that the project sequence is appropriately placed in the environment—is done when choosing the project sequence and deciding the phases, work which has been described above. While this may be the main visible activity, the feasibility team will have to keep returning to these areas to make sure all bases are covered. An example of the need to maintain an overview to ensure the project sequence is appropriately placed in the environment might be where there is a competing project and one needs to develop some strategy to fend off or defeat the competing project. We will come to this later when discussing the parallel feasibility phase.

Managing the phase

Someone is required to do the work of the feasibility phase. People have to be found and their work managed. To do this, it is recommended that the first step is to appoint a project manager, who then selects the feasibility team and conducts project start-up meetings. The project manager also manages the ongoing work of the team by concentrating on the development of an internally consistent and agreed project story and by applying the control framework.

The feasibility phase is probably the only part of the project that will engage the project manager in activities close to the actual project, that is, it will give a sense of being at the coalface. In later phases, the project manager will be much further away from the coalface and will have to make a deliberate mental effort to keep at a distance.

The feasibility study itself can be a project, with management going through the phases of planning, doing and handing over the phase (note the phase itself does not have phases, the management work has the phases). Remember that the feasibility phase is managed by a project management sequence and its project is the Project Proposal.

Appointing the project manager

The project advocate will have to push for the appointment of the project manager. Once the project has got to a stage where the need for a feasibility study is recognised, it is possible that many people will emerge with an interest in appointing a project manager, to advance their view of the project. Thus, the last fundamentally political act of the project advocate is to influence the appointment of the project manager to ensure the project takes the desired form.

Internal or external project manager

Some client organisations appoint management consultants specialising in project management as their project manager. The employees of these companies are in the same mode of 'Have gun will travel' except it is 'Can run project will travel'. They identify themselves in a professional way with the interests of their client (who may be the project client or the project supplier) and act accordingly.

Appointing a project manager from an outside organisation is a serious decision. However, it is a reasonable one if the project is outside the company's core business, the company is stretched, or the company wants to learn the skills of project management by allowing its own staff to work under the outsider.

Characteristics of the project manager

A great deal of research has been done into the traits of good project managers. In Chapter 1, some of the desirable characteristics were discussed; these might be considered when considering an appointment. Remember that at this point the appointment is of a project manager for the feasibility phase, not a project manager for the whole project. Whoever is appointed to run the feasibility phase probably needs to be a person who can live with, and possibly enjoy, ambiguity and uncertainty.

There are no easy ways to identify and select the best project manager when there is a range of candidates. It is an area of risk and the client can only reduce that risk by either taking care that good teams are selected (as against putting too much emphasis on the project manager) or closely monitoring certain critical performance indicators to give warning of potential difficulties—or the client can do both.

Appointing the feasibility study team

The team can be appointed directly by the client or indirectly through the project manager. It would be usual for the project manager to identify the team and make recommendations to the client; formally, the team would be contracted to the client. Professional appointments are often made on a cost-plus basis. In finding the team members, personal recommendations can be followed up or appointments can be made by open invitation.

This is the first time, but it will not be the last, that the project manager has to employ experts with knowledge he or she does not possess. Not only will the project manager have to find these people, and that means knowing who the good ones are, he or she will later have to decide whether or not to rely on their advice.

In many industries, there is in fact a standard set of experts. In the building industry, there are architects, structural engineers, electrical engineers, mechanical engineers, electronic engineers, cost engineers and so on. In the computer software industry, there are system analysts of various types; in the health area, there is a very wide range of specialists. What is interesting about these divisions of expertise is that there is usually a long history associated with each expertise and well-laid out terms and conditions of employment.

What the author is beginning to notice is that because each project is now so technically complex, almost all the experts are required on every project.

In appointing experts, the project manager needs to outline what is generally expected of them (the project manager will, in many cases, be able to use documents from previous projects as a guide).

The document inviting proposals should describe the project in so far as the project is known, its objectives, who is involved in it, financial constraints, timeframe and so on. Remember we are only at the beginning of the feasibility phase; we do not yet really know what the project is.

Where this phase differs from the implementation phase is that what the project manager is doing is inviting *proposals* from the experts. The project manager is not defining the way the experts will work, just outlining the general expectations. It is the job of the experts to say what they propose to do—they are, after all, the experts. Their statements of what they propose to do will cover the services they propose, timeframes and associated costs.

Now the project manager has to evaluate the offers. There will be a wide range of considerations here—one factor is to be sure that the expert actually has a sense of what the project is really about. Technical competence must be evaluated, and this may have to be done by reference or by reputation. There is no certain way. For many forms of expertise, there are important evaluation criteria such as cost, depth of organisational support, reputation for timely delivery, professional presentation, degree of confidence exuded (this might be an important skill in some tight situations), and so on. All these must be considered. However, despite all the emphasis on competencies, it will be impossible to measure the really critical, central skills involved. It is impossible to identify the best designer, the best architect, the best engineer or the best organisational change agent.

This problem is heightened in project management because of the one-off nature of projects. Some experts work regularly with experts in another field—they have found each other compatible and skilful and do not want to risk an existing sound relationship. One in fact may be hiring clusters rather than individual organisations.

Another difficulty is that the best technical skill is often buried in consulting organisations that have front people to do the selling and marketing; these people are the only ones met when evaluating the offers. The project manager should keep open the option to look inside the organisation when choosing the experts.

There is also the case of the disappearing expert. When the bidding company made its offer to your project, it offered the services of *named* highly skilled people with very impressive track records. They are chosen on this basis and the work starts. Suddenly these highly skilled people have to go off to another project and less well-qualified people are offered as a replacement. It always happens. On projects where the continued services of the named experts are critical, and where one can see a way to carrying out the threat, make it a condition that the contract is nullified if the named experts are withdrawn.

Because the work in the feasibility phase is exploratory, much of the payment will be on a time basis.

Managing the feasibility team

At this point in the project sequence, the project manager will be leading a team involving professionals. Professionals usually expect to have considerable control over the way they work and often like to develop some mystique. The project manager can allow all that, but must insist on the outputs or deliverables.

Having been selected, the experts will advise the project manager on matters beyond the project manager's competence; it is a significant characteristic of projects of medium to large size that the technology of the project is beyond the competence of any one project participant. This usually presents few difficulties, particularly on common-type projects where the advice from the experts is likely to be consistent both within the set of experts chosen and also among any experts turned to for a second opinion. On unusual projects, this will not always be the case. The advice will be mostly consistent except at some key, critical points.

The management activity can be much more technically orientated in the feasibility phase than in later phases.

A useful device in managing the feasibility team is the project start-up meeting.

Project start-up meetings

At the points where the project moves from one state or phase to another, project start-up meetings are useful. They are simply meetings where the people involved in managing the project get together for a few days to go through all they know about it and to reach agreement on how they will go about the tasks of the phase. At a more subtle level, it is a device through which they come to agree on the project story and gain some cohesion as a project team.[47]

The project start-up meeting is there to give a boost to the project, to gain momentum and to raise morale. The project manager must seek to gain the commitment of the project participants. That commitment can always be measured later in terms of the amount of work and resources devoted to the project, but gaining it is vitally important.

The meeting can be used to agree the briefs or modify them in the light of what transpires. Commitments to timetables, methods of interaction and so on can be discussed and agreed. As the discussions proceed, the project manager can move things towards appropriate team structures, reporting methods and so on.

The project manager might be tempted to make all these decisions alone and not involve the participants. This would be a mistake. These are relatively safe topics and people can discuss them fairly easily and in relatively open ways. People get to know one other and start to build up trust.

In the event that a project start-up meeting cannot be held, the project manager will have to make many decisions alone and then sell them individually as the feasibility phase proceeds. Much of the team-building work that is carried out either implicitly or explicitly in project start-up meetings will have to be achieved in regular project meetings and other informal processes.

The project stories

The project stories probably started in the transition phase but now need to be developed and agreement reached on various understandings. What is crucial is that the project manager look for inconsistencies in people's understandings of what the project is and how it is being achieved.

The next phase

Finally, how to hand on the work to the next phase. The main vehicle will be the project proposal which provides briefs for the planning and design phase. This document needs to be developed with the understanding that others will have to take it up, without the informal understandings that exist among the members of the feasibility team. It therefore needs to be written in the quality management framework of dealing with a downstream customer.

The handing-over of the project proposal to the client closes the work of the feasibility phase. The project manager, however, must now identify the work to be done in the next phase of the project sequence.

Chapter Five

5

Planning and Design Phase

During the feasibility phase, the client's needs were transformed into 'Project objectives', and then further transformed into a 'Project', where that project is expected to meet some or all of the client's needs. Associated with the description of the project are estimates of time and cost. These are detailed in the project proposal document, which is a significant input into the planning and design phase.

Now the project manager must deal with the problem of getting it, of acquiring the project. A set of instructions must be produced that can be acted upon by others in the implementation stage. Typically, those instructions consist of drawings, specifications and other contractual documentation. There are two main objectives in this phase: the first relates to concept planning and design, and the second to planning and design documentation. The objectives are:

- **to identify all the parts of the project, how they work and how they function together; and**
- **to prepare instructions to which the relevant manufacturing, construction or service industries can respond in the supply of the project in the implementation phase.**

At the end of this phase, the supply industries will have extensive data which they can work with and move into the implementation phase.

There is another planning function, not described here but referred to in the previous chapter, which relates to the planning of the project sequence; this is taken up again in Chapter 9. For example, the details of the links between phases of the project sequence have to be worked out and, while these do not form part of planning and design, they both influence and are influenced by design. While the work done in the *phase* of planning and design is conducted within this planning framework, the directions to be taken by the project sequence are to be found in the planning and design of *the project*. Planning and design becomes the central issue, because it details what has to be achieved. All the other matters are then organised to bring about what is decided in planning and design. Planning and design extends beyond the detailing of physical objects, fashion, aesthetics and architecture; it includes, depending on the project, organisational design, information system design, advertising design and so on.

The work of planning and design

The planning and design phase describes the project—it describes it in such a way that people know what is required and can go about organising for its acquisition. The output or objective of this phase is the transformation of the functional definition of the project, and the more advanced definition developed during feasibility, into a statement or description that industry can respond to in supplying the components of the project. This can be stated as comprising two steps:

- giving detailed form to the project that is expected to meet the client objectives, and
- preparing instructions to which industry can respond in the supply of the project.

In the planning and design of engineering projects, output usually takes the form of drawings and specifications; on software projects, it is specifications and logic flow diagrams; on organisational change projects, documents usually describe work flow, organisational structures, job descriptions, advice to be given to certain people, course outlines for training, and so on. On some projects, however, particularly those with a political content, the full planning may be hidden from view.

The planning and design phase is about going from the definition of the project that emerges from the feasibility phase to a point at which industry can respond. The feasibility phase took the general idea from the transition phase and developed it sufficiently to make some key choices. At the beginning of this phase, people know what they want in broad terms—they want a building, a machine, a marketing strategy for a particular product, or to find someone lost at sea.

What is wanted now is the exact form the project should take, which needs to be further elaborated, further described, in sufficient detail that suppliers can respond and agree to supply. The level of detail required may surprise people inexperienced in projects. Anything new requires considerable thought—because each project is unique, it requires extensive thought and detailing. Most of this occurs during the planning and design phase.

Concept and documentation

There are two steps in the detailed consideration that takes place during this phase. The first step identifies the particular viable solution to be used, and the second provides the precise details and properly documents the particular solution. Three examples are given.

In the planning of the film, *Psycho*, Alfred Hitchcock decided that the murder scene was to occur in the shower with various other dramatic details—step one. The second step would have been the determination of the number and type of camera set-ups (and a large number of set-ups were used) needed for this short sequence.

In the design of a submarine, a whole range of systems are needed, one of which is a voice communication system. Step one is to identify the most suitable system or particular performance specification; step two is to detail it out so that it can be manufactured or purchased and installed.

An air-conditioning system needs to be incorporated into the design of a building. The choice of the particular system is done in step one, while it is detailed out in step two.

In all the above examples, the first step comes under the heading 'concept' while the second is 'documentation'. It is of vital importance to understand that these two steps are separated out only to assist in the management of the project sequence. In the examples given above, it is fairly easy to see how these steps come about. In some industries where it may be difficult to see how these separate steps exist, they should not be introduced.

There can be many repeating cycles within the planning and design phase. Take, for example, a research project to develop a new material—nylon was discovered as a result of a very specific project to find a new material. Such projects do not just go through one concept phase and then one documentation phase; in fact, they might go through many repetitions of these steps. For example, in trying to find a new material, a chemist or metallurgist might develop a theoretical idea which might work. This is part of concept and might lead to many calculations to identify how much of each type of material is needed and the sequence of mixing. This is still at the concept stage but is getting close to documentation. The next step would be to write down the instructions for the laboratory experiment—this is documentation. The laboratory experiment might then be regarded as implementation. If the experiment fails, then it is back to concepts and through the steps again. The steps are put there to assist in the management of the project and will vary from project to project.

In the engineering industries, concept design will consist of the production of drawings and specifications called schematics to be followed by full documentation. The schematics part fits into concept and the full documentation is (surprisingly!) called documentation. The schematics essentially solve the technical problems to achieve the objectives of the project. In building design, engineers will, in the concept phase, take it to the stage of specifying the thickness of the concrete slab and will then detail the reinforcement in the documentation phase. Architects solve the layout issues of a building and give indications and some details of aesthetic treatment in the concept stage (this is often presented in drawings of such a scale that the detail cannot be shown, say, 1:100). In the documentation stage, the full details of the building will be described right down to the bane of every architect's life—the door schedules.

There are both similarities and differences across industries. Despite the varying needs of different industries, and the different sequences, there will be common steps as well as unique ones. What will often happen is that the same step, the same work, will be given different names in different industries.

Concept

The activity during concept is the translation of the functional requirements of the project into a specific choice of components or elements. The objective of this stage is to identify a particular feasible solution capable of satisfying the functional requirements. (Note that there may be many competing solutions but only one can be chosen. However, there are exceptions where one might take a limited selection of solutions into the documentation stage and in that stage make a final choice.)

The output is the decision to proceed with and to pass on to the design documentation stage a specific project consisting of specific components or elements.

The most imaginative part of the planning and design sequence usually occurs during concept. Planning and design adds much value to a project: a good design can lead to an attractive new product, a well-functioning piece of machinery, a fine building or a healthy organisation. The quality of the planning and design is a concern for a project manager. Previously the design sequence was presented as a problem-solving sequence; this was done then to facilitate the work of the feasibility phase. Now, in the planning and design phase, much more design work needs to be done, so the sequence will be revisited.

Planning and designing the project

From the project manager's point of view, the key part to get right is the combination of function with the relevant artistic, cultural elements or symbols of the project. It is the artistic or cultural element that seems to bring the mystery into planning and design, and it is this element that often adds most to the value of the project. The challenge for the project manager is to get both the functional and the artistic or cultural elements right.

The word 'design' covers a huge range of activity and output. The term starts at the artistic end and moves across to quite mundane outcomes. It ranges from craft materials to the high-tech power of computers. It also covers a wide range of human skills from musical composition and choreography to the orchestration of social change. Defining the type of design required for projects is difficult; designers have not come to any agreement on its definition, which of course leads to mystification of the design process.

There is no one way to go about planning and design. People approach it in different ways and use different steps to get there. Most designers live a life where things are relaxed at the beginning of a project and then come to a crescendo of activity at the end of the design sequence. It is quite remarkable what can be produced in the last few days of a design sequence.

Functional Solution

The type of planning and design being considered on a project is goal-orientated. It is undertaken to satisfy some need, to achieve some function. People who earn their living producing solutions to meet goals include architects, industrial designers, graphic designers, fashion designers, engineers and computer programmers. Some of them operate within a technology while others see themselves as operating across technologies; some of them regard their work as art while others are not conscious of any artistic component in their work. What they all accept is that there is a purpose to their work, a need to be met and a function to be satisfied. Part of the work of this phase is establishing quite clearly the functional requirements to be met so that the designers can be properly briefed. Most of this data should be available from the project proposal. If it is not available, it must now be developed. Failure to develop such a definition is very likely to lead to project problems, or a waste of resources.

While it may not be so obvious, design skills and design work are needed to solve functional problems. It is important for the project manager to be on the look-out for a good functional solution as against a clumsy one. The first solution to a problem is very important in that it establishes that the job can be done, but it is often clumsy; usually there is much room for refinement. One example is where the first solution, that requires lots of different components, is refined to a solution that needs much fewer components, or where the solution is refined to a point where it is much easier to manufacture. Another example is where the functional design is refined such that errors are reduced—a practical example of this might be where the design of electrical connectors is taken to a point where an audible click is heard when the proper connection is achieved.[48] Oddly enough, one will hear software people talking of an elegant solution to a software problem or an elegant subroutine, as opposed to a clumsy solution or a clumsy subroutine.

For most project managers, an efficient, simple, functional solution is not in any way mysterious. Evaluating function, that is, tuning in to how well the proposed design is expected to work, is a difficult exercise, but it is clouded in less mystery than the issue of aesthetics. The project manager should always examine a proposed design from the point of view of function and call for explicit explanations as to how the proposed design satisfies the functional needs. For example, a project manager should always insist on being fully briefed by the architect on the layout and functioning of a building, should always insist that the industrial designer shows how the proposed product works, should always insist that the proposed screen layouts, developed by the system designer, work.

Artistic or cultural element

What may be a problem here for the project manager is the view that artistic or cultural work cannot be subjected to management. This may be true in relation to the great works of art but here we are concerned with a project, and that has to be managed, but with some sensitivity. The dilemma is that, while a fine solution to the aesthetic or cultural element will add enormously to the value of the project, no-one can guarantee it will emerge.

In a way, it is practically impossible for a project manager to pronounce on the artistic merit of designs. In practice, the artistic merit issue is side-stepped when the designer is appointed; the designer becomes the judge and arbiter of aesthetic standards. Psychologically, it is often necessary for the designer to maintain a quite arrogant position in the area of artistic merit—any area where uncertainty exists and one has to rely on subjective judgement is usually dominated by some form of arrogance or self-belief.

The planning and design sequence

Despite all the above, design for projects is usually carried out in a set of steps, previously described as the problem-solving sequence. Some designers would dispute this sequence, regarding design as a gradual teasing-out and evolution of the one correct proposal. The project manager needs to know which view is taken by the designer and respond appropriately. In the case of the options approach (described as the problem-solving approach), the project manager can seek details or progress towards the options, while in the case of the teasing-out, the status of the elaboration should be discussed.

Designers often talk about the idea behind the design. This is often open to discussion and usually emerges in relation to the functional requirement. The idea is often related to emotional and intellectual responses that are being sought from users, to ways these responses may be induced, and so on. There may be a very flimsy apparent relationship between the ideas and what is actually there in physical form, but the idea is still a vehicle.

The designers start work on the problem, collecting data and looking at the ways others have tackled similar issues. Ideas begin to emerge, ideas which might be taken into the design solution. The designers may go through a disaggregated phase where they take individual bits of the problem and produce responses to them. An organisational design consultant will spend much time talking to people, reading organisational documents, identifying key players and so on. The results may be sketches, three-dimensional models or written plans of action. Having mulled over the problem, the designers can then usually see one or more ways of tackling the design.

As with the problem-solving process, the step of identifying the design always seems to involve a leap. There always seems to be a leap to a solution; there does not appear to be any step-by-step sequence to take one there. Having taken the leap, the designer evaluates the solution, checks it to see if it works, and generally tests it. Through this process, a designer may go through many many possibilities, dismissing the ones that don't work and continuing with those that offer promise. But eventually, usually very soon, the designer has to decide on a small number of options, then move into a sequence to identify the solution to run with on the project. Some designers like to quickly dismiss options and put the effort into the most promising solution, while others like to keep as many options open as possible.

Design of the product and of the production process

On many projects, not only does the product have to be designed, but also the method of manufacture. Where the method of manufacture is already available in industry, this is unnecessary.

In bringing a new product to market, both product and method of manufacture need to be designed. During the design process, the manufacturing process will be taken account of. The development of a new product, therefore, has this interesting dual design problem. As regards the product, aesthetic values will come into play but will not be a significant part of the design of the production process (although there are production processes where the aesthetics have clearly been considered).

On building projects, these production issues exist but they are mostly solved already—the steel industry supplies the steel parts, the concrete industry supplies the concrete, a brick industry, a timber industry, and so on. Designing in these industries is usually conducted within the capacity of these supply industries to deliver. In relation to the introduction of the new product, the production process is largely designed within the capacity of the relevant component industries.

Prototyping

A design method which should be mentioned is prototyping. Prototyping is a process where one decides to look to practical experience as an important guide for the design. When the design process reaches a particular point, a point where sufficient information is there to allow an attempt at the solution, a prototype might be constructed. This is quite viable in the development of products such as white goods, motor cars, software and items of mass production. It is probably quite viable in the design and development of some special large one-offs, such as in a space program, but otherwise it is limited to testing parts of a larger project. Prototyping is limited by the size of the project (a prototype of the Alaska pipeline is not appropriate, but a prototype of a welding sequence is). It is probably true to say that prototyping is underutilised in planning and design.

Presenting the output of concept planning and design

The work of the concept phase is presented in a wide variety of forms, some of it quite flimsy and insubstantial, such as butter-paper.

Documentation

Like the feasibility phase, the concept stage suffers because it is an early stage where information is little more than minimal. At the end of concept, however, considerable detail is available to be fed into a documentation system to record and clarify the data and information. Usually, the documentation is produced under the control of the planning and design professionals, who normally produce it in a form relevant to the needs of their particular profession. Much of this output is not be in a form directly usable by the people in the supply industry, who will later have to use it. So the suppliers reorganise the data into a form they can use to produce or supply what is required; this is done in the pre-implementation phase.

The objective of documentation is to detail out the decisions of the concept stage so that the supply industry can respond. In some cases, it requires making quite detailed documents and instructions, particularly if the item is quite different from the usual. In other cases, the detail need only be slight, relying on the experience and know-how of the supply industry to produce what is required.

On building projects, there are drawings showing the overall arrangement of the project and how things fit together. There are also sets of drawings showing the work of the technical specialist consultants—the air-conditioning consultant will detail the work in air-conditioning drawings and so on. Sometimes there is an exact overlap between the consultants' expertise and the subcontractors who respond in the implementation phase, for instance, with air-conditioning subcontractors. At other times there is no overlap, and one of the jobs to be done in the pre-implementation phase involves disaggregating the data in the drawings and reassembling them in subcontractor packages. An example of this is the production of workshop drawings for steelwork.

The idea of work packages emerges at this point. This will be dealt with when the pre-implementation phase is discussed in Chapter 6.

Again note that in the transformation of the functional definition of the project into a description to which the supply industries can respond, there is the possibility of the description getting out of line with the project objectives. This has to be handled by the management of the planning and design phase and the overall management of the project sequence.

Working towards handover

An aspect that is often overlooked is that this phase also has to lay down the rules, the conditions, the methods by which the work of the supply industries will be considered acceptable. The planning and design phase has to determine the means by which successful completion of the project work has been achieved and identified. In everyone's mind, the point where the project is accepted as complete is so far away that it is hardly worth worrying about at this stage. While all of the work defining acceptance may possibly not need be done in this phase, much of it has to be done. The sequence of commissioning and handover actually needs to start in the planning and design phase.

A characteristic of projects is that they have defined objectives by which completion can be identified—what those defined objectives are should be clear from the documentation. The answer to the question 'How will it be known when the project has been completed?' must be clear before concept starts; if it is not clear, then it needs to be clarified as a first step. As the concept stage unfolds and as documentation is developed, the completion of the project has to be further elaborated, elaborated to the point where someone can actually decide that it is completed, and knows how to decide that it is completed. How to know that the machinery works, how to know that the software functions as expected, how to know that human resource policies have been properly implemented are key questions to be answered before the client takes over the project from the supplier.

The documentation produced at the end of the planning and design phase must include procedures for the acceptance of the project. An 'acceptance test specification', for instance, has to be produced in the process of managing hardware and software projects. This form of documentation requires expertise in its preparation; the planning and design consultants are the appropriate people to do it.

A dilemma for suppliers can arise where the client does not get actively involved in defining the product requirements or refuses to clarify objectives. Where this happens, the supplier can have difficulties and needs to take steps to avoid difficulties in the handover stage. The contract will have some guidance, but if not there may be problems in getting paid at the end of the contract.

Preparing the documentation

The usual planning and design documentation includes drawings, specifications, networks, breakdown structures and budgets. Documentation covers a wide range of media but mainly it is either on paper of on electronic media. Each of those media allow a wide range of presentation forms—with multimedia, the possibilities have suddenly exploded.

With the advent of electronic documentation, new possibilities have emerged but also new difficulties and new dangers. Inconsistent documentation is probably easier now than previously; some forms of documentation are more stable than others and so on.

The documentation will always be a growing and developing element. It usually goes through several stages from draft to final copy. For some documentation, there are many stages: there are initial documents, developed documents, documents with some commitments associated with them, and finally contractual documentation. Documentation often bears a range of signatures, where each signature indicates different types of responsibility—some signatures indicate technical responsibility, while others indicate commercial responsibility.

Invariably, documentation work involves spending a lot of effort checking (important but in many ways unrewarding work).

Documentation should be prepared in a way that allows traceability, that is, one should be able to find the data or information on which the document is based.

Managing the work of planning and design

We now turn to the management of the design work and the management of the planning work. Previously, the work of design and the work of planning was identified; the focus of this work has been the project itself and the project sequence. But now, the planning work and the design work must be managed.

Management of the work of planning and design requires that the planning and design organisation be set up, and then that organisation be managed on an ongoing basis.

Designing within the capability of industry

A few quality management ideas should be considered in guiding the design effort. These ideas help to focus on what industry is capable of producing, thus enhancing the possibility of a successful outcome.

Quality of product

In very crude terms, this relates to deciding whether or not to produce a top-of-the-line item (the Rolls Royce) or an item less ambitious. In most product categories, there are divisions commonly regarded as quality divisions. In fact, marketing exploits this and many companies make strenuous efforts to be clearly seen in one of these quality divisions. The sequence of choosing the product can employ a range of methods, such as 'House of Quality' and 'brainstorming'. Design must always successfully deal with the quality of product issue.[49]

When applying the quality of product idea to projects, that is, to the identification of projects, one has to use stakeholder analysis and other techniques that identify the interests, needs and wants to be served.[50, 51] Very often, this actually happens in the feasibility phase, and it must certainly be elaborated on and completed by early in the planning and design phase.

Threshold of complaint

We now turn to the quality of sequence or quality of process issue. In order to make specific targets achievable, they need to be specified as a range, not as a precise number. In the physical world of production, one cannot produce to a precise number.

The target is set in relation to the customer's expectations. The threshold of complaint is a measurement of some feature which affects quality. It is the point or state beyond which the satisfaction of the customer is not affected. Another way of viewing it is to say that it is the borderline between 'okay' and 'not okay'. The threshold of complaint implies a range (note that it is a range and not a point) within which what is done is considered satisfactory or correct; outside the range it is unsatisfactory or incorrect.[52]

Thresholds of complaint can be quite extensive, covering such things as cost, time, quality (ie, as specified), dimensions, durability, stability (eg, of software code), morale (eg, in an organisational change project), symbols (forms or trappings of success), user documentation and specified legal requirements. Finding a measurable sense of the threshold of complaint on projects requires some innovative thinking.

On many projects, the threshold of complaint is implied in specifications, but the steps to take it through to the point where it is stated as a threshold of complaint are not carried out.

Setting up thresholds of complaint forces one to think in positive terms of what is wanted or required. One is forced to state what is wanted (positive) as against stating what is not wanted (negative). This positive framework is difficult to establish and maintain; some attempt will be made to look at the pre-implementation phase in these positive terms (a slip into negativity will be required now and again).

Capability

Capability is a measure of what can reliably be achieved or produced by a production effort. Under certain mathematical conditions, that is, when a process is in a state of statistical control, the output from the process is reliable and definable. These mathematical conditions can be precisely defined and achieved in relation to repeating, ie, process, activities but are not easily defined in relation to one-off activities, to a sequence. Thus the application of the idea of capability as developed in quality management needs some modification in order to apply it to project management.

Instead of requiring that the sequence be in statistical control, the idea of 'sequence capability' will be a measure of the expected output from any sequence. For example, there will be an expected completion date—whether or not this is the required completion date is another matter. The measure is mathematically difficult, probably involving both a range of values and a set of probability distributions. We will only use it conceptually—this can be useful in that it can provide some pointers to managerial activity. Every sequence of activities has a capability. Every surgeon can make an incision, but some more precisely than others; almost any piece of machinery can produce an item within a very close tolerance around a particular value.

In calculating capability, consideration should not be confined to thinking in terms of simple physical measurements; issues such as adaptability—for instance, how well can subcontractors adjust to change and delays—need to be addressed.

Ability to self-correct

An important idea in quality management is the idea that people doing the work ought to be able to make adjustments as they go along. They ought to have their target range and take measurements as they proceed, and thus be able to take corrective action. If the sequence of work does not allow for corrective action, much failure can be expected.

On projects, there are processes in the implementation phase that repeat quite often and are therefore quite closely related to manufacturing. To these processes can be applied fairly standard quality management techniques based on repeating processes. A key part of these techniques is monitoring or measuring the output as work proceeds. On the basis of output measurements inferences can be drawn on the state of the production process.[53]

Matching threshold of complaint and capability

In practice, requirements should be adjusted according to what industry is capable of delivering. Good project planners and designers take account of—and take advantage of—this capability and alter their plans and designs so that what is specified can be delivered. Many other planners and designers operate within the capability of industry without being aware of doing so and thus fail to take any special advantage; others are unaware that their plans and designs are unworkable, which leads to lots of problems, problems the project manager has to face later.

This is a major quality issue. In some instances, whole industries live with fictional views of what can be delivered. An example of this fictional view is the extent to which the reinforced concrete industry fails to come to terms with difficulties encountered in placing reinforcing steel accurately. It continues to believe that reinforcement can be placed much more accurately that is actually the case. The result is inadequate protection for steel with consequent corrosion. Another example is the optimism of the software industry in its capacity to deliver to deadlines. Many define thresholds of complaint that are simply out of the reach of the supplying industry or have no meaning even to those defining what is required.

Since it can easily cause problems in the implementation phase, the project manager must be aware of the fact that many project industries have unrealistic expectations of their capacity to deliver. A wise precaution is to make sure the planners and designers take account of industry capability during planning and design; the project manager can then review the documentation of this phase and decide whether or not the specifications are within industry capability.

Setting up the planning and design organisation

The people who will work on the planning and design phase have to be identified and organised.

Formal and informal structures

Whenever the project manager sets up a formal organisation, an informal one will automatically arise. Much of the work that is done on projects relies on what is known as the informal structure; the work actually gets done because people are committed to doing it. The project manager would be well advised to listen out for and tune into the informal organisation—some of the most important decisions will be taken here and then be transferred to the formal organisation for ratification.

Two levels of symbolism

What is very interesting about the planning and design phase of the project is that there are two levels of symbolism—there is the symbolism that is needed to manage the ongoing phase itself and there is the design activity which is involved in the manipulation of symbols. As the project proceeds through the design phase, a phase of great uncertainty and a phase whose job it is to find and solve problems, it will be necessary for the project manager to engage in symbolic activity, particularly activity that signals that all is under control and going well (although no project is ever fully under control and is rarely going well in this phase).

Some of the work of planning and design is itself deeply rooted in cultural symbolism, which the designers use as a way of keeping the project manager away from managing the ongoing design work—how could a mere project manager know about the use of such fine cultural elements? The key to managing this is to learn their vocabulary, the words that identify the symbols.

Professional and technical skills

Staffing the project consists of filling in its formal structure, a temporary structure usually bound together by contracts. The project manager now has to decide the structure of the contractual relationships, which means deciding with whom to contract for the planning and design services.

The project manager has to identify the relevant profession or technical skills that the project needs, has to specify the expectations of those who possess those skills, the professionals, and select and monitor their work. The interfaces between professional disciplines must also be managed, and the involvement of the professionals finalised (which includes deciding whether or not their work is satisfactory).

The trouble with all of this is that the skills of the professionals are all, except usually one, outside the skill-base of the project manager.[54]

When the formal structure is in place, the informal structure will start to develop and either contribute to or detract from the project. There is usually goodwill initially, so the informal structure should, at least at the beginning, contribute. Project start-up meetings are very useful to cement in place a compatible and supportive informal structure.[55]

Representatives of the client

Before deciding on matters on the supplier side of the project sequence, it is necessary to consider the client. The client has to make a wide range of very difficult decisions with quite far-reaching consequences as the project planning and design proceeds. Because the decisions have such an impact, a real threat to the project is that no-one on the client side will make them. This failure is often based on the belief that if one does nothing, then one cannot be blamed.

It is vital that a client representative or representatives be appointed. The project control group is very valuable here (at least, it can be used to spread the blame)—it will usually make a decision, if there is a recommendation before it.

Someone on the client side must, therefore, prepare recommendations; that should be the function of the client's project manager. If the client's project manager fails to formulate the recommendations, the suppliers may have to do so (but in a manner that does not leave them open to charges of professional negligence).

The tasks to be done

This step is almost inseparable from the next one of identifying the skills required, but it is useful to think through exactly what it is that must be done. Tasks are the work to be done; skills are the ability to do the tasks.

The design and planning skills required

Practitioners in any one industry know how they are organised, and they treat their way of organising themselves as ordained by nature, not as determined by social forces interacting with technology. Some skills are common across industries, such as accounting, while others are very specific, such as programming in the software industry.

Each industry has organised itself in a particular way; certain people are expected to do certain jobs. In the construction industry are architects, engineers, quantity surveyors, builders, tradespeople and real estate agents. In the film business are producers, directors, photography, sound, financiers and, of course, actors (one only has to look at the credits for a film to become aware of the range).

The project manager has to identify the industries that will contribute skills to the project and, to do this, might try to site the project within some industrial framework. The project may exist simply within one industry or cross industrial boundaries. If the project involves the construction industry, the computer industry and the organisational development industry, then the project crosses normal industrial boundaries. Most modern projects do cross boundaries and so are supplied by a range of planning and design skills chosen from a range of industries. In practice, while project managers have to select skills from a range of industries, it is often helpful if a lead industry can be chosen.

Finding appropriate staff

The project manager not familiar with the way an industry is organised needs to talk to some people from within that industry. Having worked out which industries are involved and having spoken to people from within them, the project manager can identify the people to perform the tasks required.

Lead planner or designer

Most of the work in the planning and design phase will require the co-ordination of various planning and design consultants. Some of these are in the same office but most are spread throughout various organisations, sometimes meeting regularly, sometimes intermittently. This dispersed network of interacting design decisions needs to be managed and controlled by a system that is visible and which has considerable authority.

The question now arises as to who integrates the work of the consultants and their decisions. It is not the project manager's job. Integrating the planning and design input from various consultants is itself a specialist function, therefore in most projects a lead planner or designer is required, a designer who decides or recommends where the work of other designers fits in. Obvious examples are the architect who integrates the work of the engineering consultants, the systems designer who integrates the work of various analysts, and the industrial product designer who integrates the work of various consultants.

It is the project manager's role to have a sequence in place that ensures the design recommendations actually do fit together; however, it is not the project manager's job to fit them together. If the lead designer is a very competent designer, then that person will ensure that things do fit together. Otherwise the project manager has to take steps, mainly by means of design reviews and configuration management, to ensure that everything works.

As can be seen, there is now a very fine line between the role of the project manager and that of the lead designer. Treading that line, the line between designer skills and management skills, the line between designer activity and management activity, between design authority and management authority, is a very skilful, subtle activity which requires considerable cultural capital. It is vital that the project manager does not assume the specialist consultant planning or design function.

In essence, the project manager should provide managerial support to the lead designer. Not many project managers really understand this role of management providing support to the lead planner or designer. In practice, the lead planner or designer often fails to achieve the integration of the specialist skills; when their failure becomes evident, the project manager steps in.

So while the formal contractual structure may show a flat layer of specialist consultants, there should be one specialist planning and/or design consultant who directs the others. By way of example, in making a film, the director carries out this role, integrating a wide range of specialist contributors; the producer, who in many cases is a project manager, has to facilitate the director's work, but not do it.

Sources of staff: internal and external

Sources of staff for this phase can be internal or external, or both. Acquiring internal resources, however, requires some political skills. Though much of the discussion on the use of internal resources for projects leads to the matrix structure, here we are concerned with more immediate needs.

To get the internal support, the project manager has to carry out an internal political campaign. This will involve 'managing up' and 'managing across'. When relying on internal sources, the acid test is always 'Are the promised resources being provided?'—with internal arrangements, where no legally enforceable contractual arrangements exist, there is always the danger of other priorities taking resources away.[56]

Getting outside consultants usually involves some formal selection process which involves preparing a brief, setting out the selection criteria, finding people willing to make offers, evaluating the offers, and then appointing the consultant with the most attractive offer.

The selection criteria might need rather careful consideration; capacity to do the volume of the work is now clearly an issue. This capacity may be indicated by sizes of bidding organisations, by reputation and by price (it is often useful to reserve the right not to have to accept the lowest price).

The general rule is that, where possible, get a fixed price for the work. The more the work is clearly specified and the more that one can expect the bidders to understand the extent of the work required, then the more one should tend towards calling for fixed price offers. The opposite conditions tend to move the bidding towards rates, usually called 'cost plus'. It is not always possible in this area of work but many companies in planning and design are able to set realistic budgets for their work. The project manager must try not to change the scope of the work and thereby cause changes to the fees. In practice, however, there will be changes of scope and offers from consultants should include procedures for handling such changes.

Establishing structures and relationships

The formal structure not only consists of the relationships set up by contracts with external sources of skills but also of understandings in relation to internal sources of skills. The formal structure has to recognise that, while there are formal, contractual relationships, practical reporting and communication relationships may and probably will be different. While some of these can be spelled out, one has to rely on the informal structure to actually make things work. Much of how the informal structure works is already part of industry practice.

In the planning and design phase, the formal structure usually consists of the project manager managing a number of specialist consultants who are contracted to the client. The consultants can range across the whole gamut of planning and design, including such areas as finance and law. The most common formal structure resembles a functional hierarchy, focused on a single project, within which are linked specialist consultant organisations. How the consultants are internally organised is not the direct responsibility of the client project manager, though it is a matter of interest and concern.

The consultants' staff

Although the project manager has no direct, formal contact with the teams within the consultants' organisations, contact should be made with them in an open, informal manner. The project manager needs to tread carefully here if continuous contact is expected; the principals of the consultants usually do not like staff talking to outsiders, particularly outsiders like project managers. Weaknesses may be exposed or commitments given.

The project manager must behave in a very principled way, using the contact to help build commitment to the project and not to undermine the principals. If, as a result of information learned during this informal process of contact, the principals need to be dealt with, the project manager needs to be very skilful politically. For instance, the source of the information might have to be protected and the 'face' of the principals protected.

One purpose of the project manager's contact with the consultants' staff is to develop commitment to the success of the project. This can be done by providing information, by asking people how they feel about the project and then dealing with the issues that arise. Achieving commitment involves the sharing of emotions and feelings.

The consultants' internal management

Having been appointed, the consultants or specialists get involved in the project sequence. They need to review the scope of the project and the data provided so as to decide how to address the issues and problems. As already indicated, the consultants usually go through the two stages of 'concept' and 'development'.

The consultants have to manage the work as a project within their own organisations, following procedures similar to those followed by any project manager. They have to identify the needs of their client, review the brief, plan the work, set up the team to carry out the work, manage the team while it's carrying out the work, and then disband it.

Typically they manage each job within the organisation by a job or project number. They set up internal documentation systems relevant to their own needs as expert consultants. If their documentation includes drawings, there will be drawing schedules. In setting up document numbering systems, they usually have to have some idea of the nature and extent of the documentation. The consultants are usually in familiar territory, which allows them to introduce systems. Thus the project manager can ask them about their work methods, their quality assurance methods, their procedures, and take account of them in the project planning sequence. Recognising these procedures allows the project manager to set markers on progress of work through the specialists' offices.

In terms of what the consultants are producing, it is clear that while many behave as if project documentation is easily understood by all, there are many working with documentation they do not understand. This is a serious problem, which often has to be sorted out in the pre-implementation phase.

While much is made of working in teams on projects, most of the teams in the planning and design phase are working from the offices of the consultants and are therefore not under the control of the project manager.

Supplier involvement

There are clearly advantages in having suppliers involved in the design sequence. They can bring specialist practical knowledge to the situation as well as insights on some particular aspects, such as the availability of particular machinery. But how does one select the supplier at this early stage?

In practice, one can only do it on the basis of setting up long-term relationships with suppliers, which implies some experience in projects, or by selecting a small number of suppliers and involving them on the basis that the choice will be made from one of their number. A possible way is to also pay these suppliers for their advice on a professional basis in the expectation that this will lead them to spend resources building up a good offer. This has variable results in practice: some suppliers pocket the fee but put little effort into the preparation of their tender while others respond well.

Involving the supplier and setting up long-term relationships is one of the key tenets of quality management.

Dealing with ongoing management

The work of the planning and design phase involves managing the performance of the individual consultants, the interface between the consultants and various external interested parties, all the time with an eye on the target—a completed project. It is the project manager who must keep the eye on the target, something very easy to forget in the sea of detail, and who must always devote specific time to take an overview.

The project manager needs to negotiate boundaries between consultants, ensure that work which must cross the boundary actually crosses the boundary, and also have ways to check that work. It is here that quality management ideas associated with the concept of the 'downstream customer' apply. While the consultants or specialists will not have an in-built concern with the downstream customer, such concern must be fostered, monitored and rewarded by the project manager.

Meetings and project start-up

Project sequences involve many meetings. Once the project consultants have been chosen, the project manager must go about welding these disparate organisations together. One good method for starting the process is by what are known as project start-up meetings.[57] Since a new project is involved and since the consultants know they have to work together, there will be considerable willingness to participate. There should be a project start-up meeting at the beginning of concept and at the beginning of documentation. These meetings may be facilitated by a professional facilitator rather than by the project manager. The facilitator, of course, should run the meeting in such a way as to enhance the standing of the project manager.

Project planning and design integration require the concurrent contribution by two or more consultants, hence the importance of regular meetings in project management. However, the author suspects that meetings serve a much more important function beyond that of integration—meetings help to keep people informed about what is going on, and help to build and maintain morale.

Meetings also put a certain pressure on the consultants to perform: this is peer pressure, a significant element that can be used constructively by the project manager. No consultant enjoys fronting up to a meeting without having the required information, or having to admit that the information will be late.

Beware of the 'I'm late because of somebody else' variety of defence: laying blame elsewhere is a good defence for non-performance! If a consultant is put in the position where they cannot or are not performing, then they may try to blame someone else. This may take the form of claiming data is not being made available from another consultant or from the client. Given the degree of uncertainty that exists in the planning and design phase of a project, there is ample opportunity to set up confusion and hence avoid censure.

So the project manager should carefully use peer pressure to achieve performance but will probably have to work hard outside the meeting with the non-performer to avoid things being sunk and poisoned by accusations of blame. The decision whether or not to terminate services may also have to be taken.

Meetings are called to help in decision-making. These are not democratic meetings where things are resolved by a vote; they are processes by which data is exchanged, consequences highlighted, commitments given by each consultant, and where each consultant maintains an essential separateness. Each consultant in effect makes their own recommendations to the client which are accepted or rejected by the client project manager.

The reader may well regard this explanation of project meetings as quite inadequate. For such a meeting to function, there must be considerable prior agreement as to why people are there, and considerable agreement as to where the project is going. The project story is one explanation of how meetings function; people already have an understanding of what is going on and what is expected—they have a story they believe in.[58]

Clearly, there is much more going on than a simple discussion of the technology of the project. It is important to recognise that when a meeting becomes difficult, frustrating and tiresome, it may be because there are actual project or project team problems rather than a bad meeting procedure. The project meeting is in the arena of group dynamics and needs to be treated as a social system with its emotional aspects. A lot is at stake at a project meeting—status, self-image, costly obligations and much more.

The purpose or purposes of the meeting should be clear, agendas used, minutes quickly recorded and given to participants (preferably straight after the meeting by the use of action sheets), and the process of the meeting kept lively and moving forward. Meetings should run to a timetable.

Re-examination of objectives and making trade-offs

The objectives developed during the feasibility phase were translated into a project, which is now the proposed project. This project needs to be examined to confirm that it still satisfies the objectives of the client and stakeholders, that it is a coherent proposal, that it still has support, and that agreement continues to exist.

As the work of this phase continues, the first of a long line of question marks arises. Problems begin to surface in the match between the objectives of the various participants and the capability of the proposed project to meet those objectives. Assumed markets may disappear, technical constraints arise, or straight political opposition emerges to some aspects of the project. These need to be managed both in the design work and in the planning work, and test the project manager's ability in maintaining clarity of purpose, an essential to project success.

The constraints on the project lead to the need to make trade-offs; it is during this phase that the major trade-offs have to be made. Almost all projects will demand trade-offs and compromises, and these should not be regarded as a failure at the feasibility phase. It is quite natural for people to have great expectations only to find that resources are limited and choices have to be made. Trade-offs are choices, choices from a set of feasible alternatives. The work of managing the phase will need to identify a clear process for the management of trade-offs; issues that can be handled within design might be done within the framework of the management of scope. But not all project trade-offs are design trade-offs; a risk trade-off might be the issue in relation to contracts, so other processes must also exist to handle changes in planning.

The big risk is that confusion might develop and then all sorts of other work is delayed and other problems emerge. Work, over which there is no question mark, will slow, and there will be follow-on effects, consequent delays and so on. Clarity has benefit in that it helps to keep work going on, but it also has a price in that work may have to be undone. As a stabilising rule, the project manager should always maintain that the approved work to date is the plan and that nothing has changed; nothing will be considered to have changed until the matter goes through the required process. People will find it quite reassuring to know that changes are to be managed on the project, not introduced at whim.

Budget control and cost contingency

The cost management process involves setting up a budget in the feasibility phase and then managing to that budget; now the process is directed towards producing plans and designs where the cost is below the budget.

The process of managing to the budget is fairly straightforward—it involves using the control framework and applying it to the design process. It has some weak components but in general the method does provide a disciplined set of steps which usually give some sense of control.

Controlling cost in the design process

Figure 3 overleaf outlines the design cost-control process, a decision-making process whereby the project is divided into elements or parts. The budget is distributed across the elements—each element thus has a cost target. There is also a contingency amount in the budget.

As the elements are being detailed out, their costs should be monitored and controlled. An element is chosen and its cost estimated. If its estimated cost equals the budgeted amount, the process moves on to the next element. If the estimated cost is below the budgeted amount, a saving can be moved to the contingency, with the budgeted amount reduced. If the estimated cost of the element exceeds the budgeted amount, there is a problem. If the cost of the component is over-budget, the cost-control function is applied, first of all to the component itself. If this fails to achieve the desired saving, other areas should be examined for areas where savings can be made. This may involve other elements being redesigned.

The problem is dealt with in the following order of priority:

- redesign element to reduce cost
- reduce scope of element to reduce cost
- find savings in other elements and transfer them to this element
- take money from contingency
- increase the budget

The first cost saving should be made, if possible, at the element level itself. If this cannot be done, other elements can be examined to see if they can release the funds needed. Only when funds cannot be found within the elements should the contingency be turned to. The project manager should put in place a formal process of approval of funds into and out of contingency; when funds are moved out of contingency, the project manager should always be notified.

The above order of priorities can be modified. For example, savings can be sought in other elements before the decision is taken to reduce the scope of the element.

FIGURE 3
Design cost control process

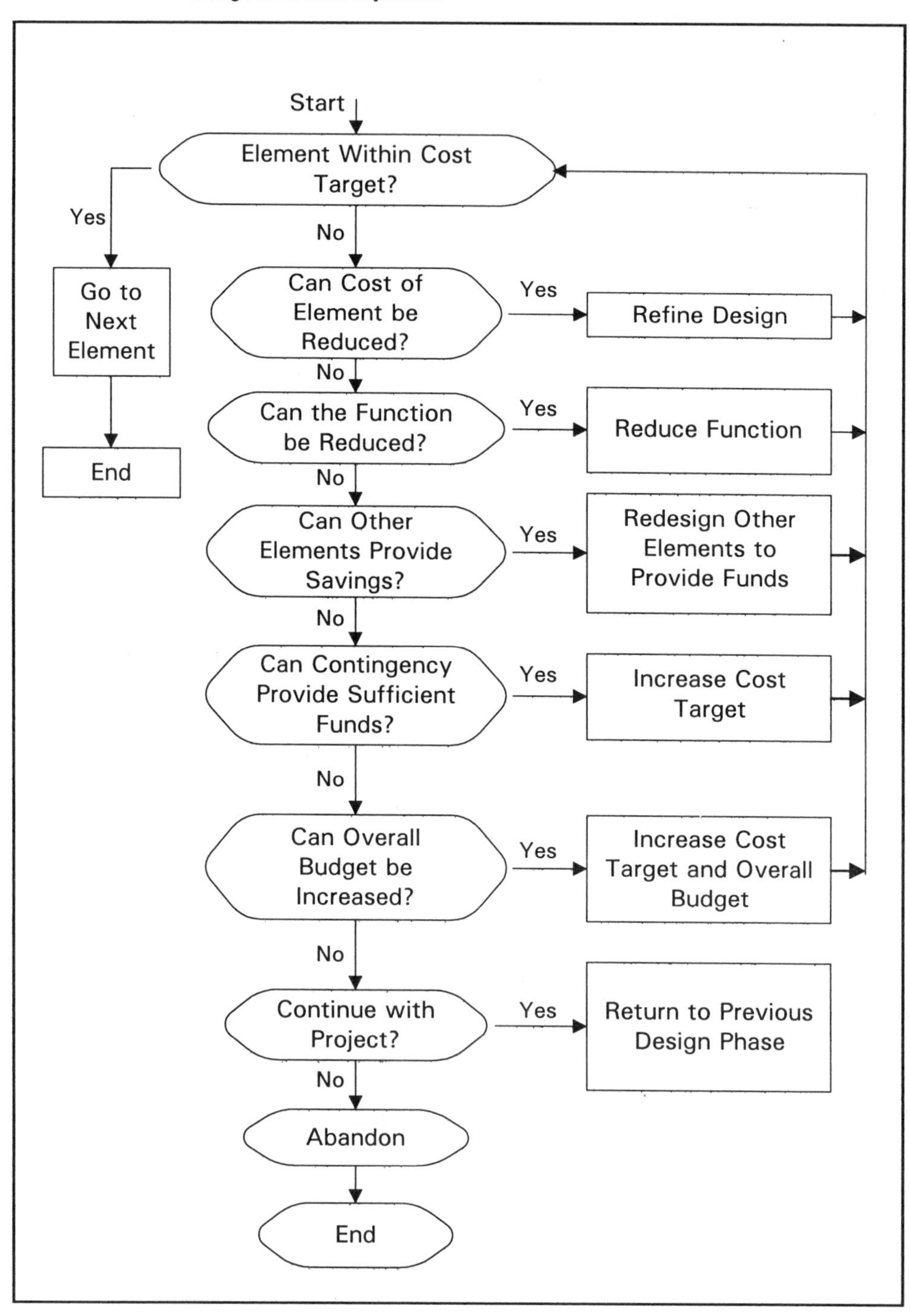

Budget and contingency

The important point is that there is an overall, total budget for the project that must be defended. It has to be accepted that there will be parts that are over budget and others that are under budget. By balancing them out, by carefully husbanding contingency, and by putting pressure on the design function to perform to clear goals, the overall budget can be defended.

To properly manage this process, it is essential to know what the appropriate level of contingency is at this stage. Although the contingency may have funds, the project may have gone over budget.

Estimating

The success of this design/cost methodology relies on an estimating system. Estimating methods are open to questioning, so to maintain consistency, alternative methods should be used to check the main method. The planning and design phase provides much data for estimating.

The method outlined helps reduce the risk of running over budget. The best defence is early warning of a problem; this means that as the design progresses, one keeps on re-estimating the costs, makes trade-offs and carefully manages the contingency. Re-estimating the cost is essential to keep track of the developing cost structure and also to guide design decisions.

Project specifications

Essentially the problem is to ensure that the specifications are appropriate, that standards are appropriate and not excessive, and that they are changed in an orderly and controlled fashion. Specifications need to be examined, particularly in those areas that seem to influence design decisions. Often, requirements in specifications are not absolutely necessary or are unduly restrictive (for instance, minimum criteria for radii in railway design have dramatic effects on positions of buildings; the value of these criteria may need to be questioned when buildings are being forced onto difficult locations). The project manager needs to appoint someone to review the specifications, to identify the key areas, and to report on their suitability.

Specifications are often there simply to protect people should something go wrong and have little real use in achieving the project. They are usually long, densely written and in small print. The construction industry is full of subcontractors who sign contracts with little real understanding of the specifications (they simply don't read them). Such specifications do not help get the job done, and they probably have hidden costs associated with their use, but they provide legal (but not necessarily real) protection. There appears to be no clear answer on how complex specifications need be, but there is a problem and it may need something like partnering to break the present impasse.

A focused client

A real danger to a project manager is where a shifting set of project requirements are imposed on the project because the client is not a single entity; for example, the client may be represented by various parts of an organisation. The classic case is where different parts of the military (army, navy, air force) have to be satisfied with one piece of equipment; some regard this as an essentially impossible project because of the problems associated with an unfocused client.

If the client does not speak with one voice, then the project manager has to find a way of bringing it about. A formal process for approval of decisions needs to be set up. The process may be lengthy but there must be a way by which a final decision is recognised, a decision which is then legitimately regarded as the client decision. In practice, much work takes place before this decision is presented. Much of the work leading up to the decision will be political in nature, gradually locking people into an agreed or at least a viable position.

Boundary management

The project manager will be required to identify and manage the boundaries between the consultants and specialists and sometimes the boundaries within consultant organisations. It is worth the effort on the part of the project manager, or whoever in the project management team has the responsibility of managing the designers, to become totally familiar with the boundaries. The location of the boundaries are best found by following the information flows and the Work Breakdown Structure. Much of blame management centres around sorting out boundary arguments.

Ease of production

There is a real need to ensure that the project is reasonably easy to achieve or make. Ease of production is obviously an important issue in the management of design, but there is very little available in the form of techniques to help the project manager know whether or not it will be achieved. It seems necessary to rely on the skills of the designers. Mock-ups are clearly a way of dealing with some of the problems, and special-purpose meetings directed at the ease-of-manufacture and ease-of-use issues are another approach. In the end, however, it will rely on the flair of the designers and consultants to identify the difficulties.

Delivery capacity of consultants

This is about the ability of the whole design system to deliver a good design. In essence, while it is impossible to ensure that a design consultant delivers a good design that satisfies a whole range of senses (eg, it is aesthetically pleasing, introduces wonder and provides elegant solutions), it is possible to take on other issues of design, such as capacity to physically do the drawings, capacity to ensure mistakes are removed from documentation, capacity to meet deadlines, and so on. In other words, capacity to meet the technical procedures and sequences that are necessary (but not necessarily sufficient) to produce the plans and design for the project.

The people who can best judge these issues are the consultants themselves. They go from project to project and can examine their own performance. In the project context, where the consultants are used once and leave when the project is completed, it is more difficult to get a good understanding of their capacity to perform. (On this point, one can see the advantages of long-term partnerships between companies where those partnerships set out on a continuous improvement process.) To overcome this problem, the project manager needs to look for indicators of performance, to look for early signs of likely success or failure.

Costs associated with the planning and design teams

The principles underlying the management of the expenditure of costs by planning and design teams are similar to those applied to the whole project. These costs are quite significant, usually in the order of 5 to 15 per cent of the total project cost.

One of the factors which makes the management of these costs more difficult is that it is almost impossible to fully specify the work; it is also possible to consume a lot of planning and design services to no avail. So financial tracking of the specialists' costs and fees should be implemented and reviewed regularly.

Before agreeing to pay bills, the project manager should review the work done and check the reasonableness of the claims; a significant amount of money should be held back until the work is complete.

Managing risk

The key risks that we are concerned with here are the possibility that the design does not work, the possibility that the proposed designs take too long, and the possibility that the designs cost too much. We are concerned to reduce the risk, avoid it, or set up a sequence that can cope with poor outcomes on the assumption that insurance is not available or is inappropriate. The issue of the cost of the planning and design teams has already been dealt with.

Design will not work, a technology risk

Where the technology of the project is outside the knowledge of the project manager (which will often be true), the project manager establishes a procedure whereby possible problems can be picked up. This is usually done by the appointment of appropriate specialists and by design reviews. This situation is satisfactory when the relevant technical expertise has been identified and can notify areas of concern. In other words, it is reliable when the experts are reliable.

Design will never finish

This is a very difficult problem. Most designers are optimistic, and thus optimistic timeframes are provided. In recognition of this fact, time contingencies are built in, often quite large ones.

However, a much more serious issue may arise. It may be that the design will never finish, that it is impossible to actually complete it. In such a situation, the project manager is forced to decide whether to abandon a project at the planning and design phase. This is a very difficult matter to decide; one way of dealing with it is by way of the Slip Diagram method.[59]

Many projects have consumed millions of dollars in the planning and design phase, only to be abandoned before the phase was finished. This is a risky area for clients. By the time concerns about the viability are raised, the client has already spent quite a bit of money. As it is very difficult to accept that the money spent has been wasted, many clients are caught sending good money after bad.

Managing time

Network methods such as CPM have usually been applied at the implementation phase and they can and should be applied to the planning and design phase. But how?

Knowing how the design is progressing is quite difficult. After all, the project manager is dealing with problems outside his or her skill-base and must therefore be advised by the experts on how long the design work will take. In practice, the designers themselves are attempting to predict when they will finish and are not necessarily confident in their predictions. The project manager thus has an opportunity to help the planning and design consultants and the project at the same time.

All expert consultants in the planning and design area have a method of going about their work. Some of them refer to it as the design process, or work process, in recognition of the fact that they need to go through various stages in order to complete a piece of design or planning work. The project manager should ask the consultants to explain their design process and how they will provide the results of their work. Delivery is almost always in documentary form or a prototype. The project manager therefore finds out the milestones in the design process, some measure of expected durations, and also details of documentation to be produced.

Remembering that the planning and design sequence is interactive, the project manager will need to go through this exercise with all the consultants, carefully noting where the consultants need to interact in order to progress their work. This can then be all set up in the usual CPM formats. What we now have is a planning and design program against which the design sequence can be managed.

During planning and design, work tends to fall behind schedule (if work gets ahead of schedule, there are also problems, but of a more luxurious kind). Rarely is planning and design work finished before time. Typically, the documentation is completed by burning the candle at both ends.

Time management is the project manager's responsibility. The fact that the work is in the hands of experts, many of whom are more talented and of a higher status than the project manager, does not absolve the project manager of this responsibility. The advantages the project manager has is that very few of the experts fully realise how important time is, and none of them has access to the time plans of the others. Thus the project manager is in a structurally strong position; the effort put into managing the time issue (which will end up as an exercise in monitoring progress) in this phase pays handsome dividends.

Contingency

One can be almost certain that problems will arise, that design will be late and delays will occur on the project. When obtaining estimates of time from the various consultants on this phase, it is important to establish whether or not they contain a contingency factor. Not only will there be risks of individual designers being late, there will also be knock-on effects from one consultant to another. Time contingencies in the design area can easily be 50 per cent.

The project manager must push the planners and designers to keep to the actual time allocated, not the time plus the contingency factor, otherwise people will take the contingency as part of their normal time allowance and time will blow out even further.

Crashing the planning and design program

Crashing the program means getting more skilled people to work on the planning and design. It is usually difficult to get crashing to work well. The difficulty is that the new planners or designers need to be inducted to the project, to understand where it is going and get to a stage where they can make an input. Initially, crashing a design program leads to delays because some of the existing designers will be taken off the job to brief the new people. After this, one may—but note it is only 'may'—start to gain time.

When the decision is made to crash a program in the planning and design phase, the project manager must ensure that the consultants clearly designate the work to be done and also provide clear plans for the management of the information flow. The consultants will also need to state the effects on their information demands.

From this, one can see how useful CPM methodology is in managing the planning and design phase. In crashing, one adds resources to the critical items and also to the items next most likely to become critical.

Managing project information

The real tool the project manager has is the project management information system. Whenever people see the words 'management information system', thoughts immediately turn to computers although computers are only a small part of this system.

The crux of the problem with project information is that the data to be managed is changing both in form and volume at the same time. So whichever information system is chosen or developed, it has to be able to grow in volume and cope with changing forms of data, and it has to be able to do this quickly.

Sample of project information needs

The project information system needs to be able to cope with some or all of the following activities, and this list is by no means exhaustive. It must cope with the management of: the scope statement, trade-offs, the appointment of consultants, activity across organisational lines, risk, time, budgets, the selection of people and organisations, procedures for handling changes and variations (initiation, evaluation, approval, implementation), a clear delineation of authorities, design impact statements, clarification of misunderstandings, intersystem and intrasystem interfaces, a sign-off procedure, project specification preparation, quality assurance and quality control, procurement strategies (contract type, invitation, responses, contract assessment, negotiations, award), the forms, documents and files to be used, the project control terms and concepts to be used, liaison with the client, the client's decision-making process, external relations, the updating of the implementation strategy, insurances, the WBS, the distribution of data and information, and the process whereby the whole team learns who does what, how and when, and who coordinates.

Concept of management documentation

Management documentation covers a range of forms and storage and retrieval. The distinction between data and information is that data is usually unprocessed facts or opinions (both subjective and objective), while information is processed facts or opinions. A management information system needs to be able to gather facts and opinions and then examine and process them. The system consists of the management documentation plus other resources, including the information residing in people's brains.

There are many kinds of documents, such as contracts, specifications, reports, drawings, parts lists, computer files, disks, print-outs, photographs and minutes of meetings. These documents describe a wide range of items on a project, and not only do they describe each item, they need to describe each version of that item as it passes through the design and later phases. Keep in mind that there is often a parts numbering system as well as a document numbering system; it is very useful if they can be linked to one another.

The author suspects that the area in which the most complex documentation comes up is software development. This is because of the various forms of existence and representations of software: there is the chart showing the logic, which may be on paper or on disk or in some other form; there is the print-out showing the coding; there is the code itself on a disk or in some other medium; and there is the machine code on disk or in some other medium. Keeping all of these logically related and yet still able to manage changes requires a significant effort.

The maintenance of the management information system

Once one becomes involved in anything but the tiniest project, the project management information system has to be developed and maintained. These costs need to be budgeted for when expected project management costs are identified.

The next phase

The planning and design phase exists to prepare data which is usually to be given to others to act upon. In the process of handing on, one must be sure that the data is in a form that industry can accept.

Chapter Six

Pre-implementation Phase

The planning and design phase decided what constituted the project; now how to provide what is wanted has to be worked out. The sequence of providing what is wanted is called the implementation phase and preparing for it is the pre-implementation phase. During pre-implementation, arrangements have to be made for the relevant parts of industry to be brought together. The actual method of execution of the project has to be finalised and the facilities needed for execution must be found and booked. This is very detailed work requiring patience and care.

While the model presented here is of a phase that follows planning and design and precedes implementation, a pre-implementation phase can essentially come before any phase; it can simply be the setting-up of a subsequent phase. In fact, the setting-up of all phases requires work similar to the work that will be described in this chapter.

A key issue is the capability of the chosen method to deliver what is expected.

In order to examine the pre-implementation phase, the implementation phase needs to be discussed.

Introduction to the implementation phase

Pre-implementation is getting things ready so that, in implementation, work can start on what has been ordered or requested.

Implementation can obviously start any time—work can start straight away without any planning, without anyone being clear on what is wanted or how to get it, or without knowing the cost or time or quality required. By starting in without any preparation, what is needed gradually becomes apparent, which is one step towards getting it. This may work for extremely small projects, but not for anything substantial. Anything remotely substantial needs planning effort before the big effort of doing it starts. It needs pre-implementation work.

Determining the number of implementation phases

There can be more than one implementation phase in a project sequence. To some extent, it depends on where one puts the boundary around the project.

Sequential phases

Consider first sequential projects. An example might be film-making and film distribution. In the making and distribution of a film, there is clearly an implementation phase while the film is being shot; there is also implementation during its distribution. Some might regard the making of the film as one project, which it is, and the distribution of the film as another project, which it is. However, the project that is concerned with the success of the film includes both implementation phases, the shooting and the distribution. A similar argument arises when examining a project to introduce a new product. There is an implementation phase when the manufacturing facility for the product is under construction, and another when the product is introduced to the market. So there is a range of project sequences where one implementation phase logically follows on from another.

Externally parallel phases

Another set of project sequences has parallel implementation phases. These projects are those that consist of parallel project sequences necessary to achieve the single overall project. An example might be the introduction of a new airline service: there is the project sequence of acquiring the plane and another in the acquisition of maintenance facilities. A most amazing project in this category is mentioned by Morris—the link between the shuttle project and the space station project. Here the space station was shown to be in danger of falling from space if it had to wait for shuttle replacement to be completed.[60]

Internally parallel phases

Inside most project sequences is a set of internally parallel project sequences leading to what is seen as a single project. Inside most modern project sequences there is a physical subproject sequence, a social subproject sequence and an information management subproject sequence.

A new sports facility will require a building, an organisational structure to manage it and provide services, and a public and internal information system. There may be more component subprojects, each of which has its own implementation phase. Thus, in most project sequences, there is more than one implementation phase.

Co-ordinating the implementation phases

Not only are there sequential implementation phases, there are both internal and external parallel implementation phases. These clearly have implications for the cash flow, for the availability of project supervisory personnel and for the demand on specific skills. What is difficult about the subprojects is that each has a different project life cycle which has to be co-ordinated. In the example above of the new sports facility, the timing of the information component subproject will depend on the timing of the social component subproject, which in turn is dependent on the timing of the physical component subproject. The helpful thing is to recognise is that it is the co-ordination of their implementation phases that determines the relative alignment of their project life cycles.

It is usually the client project manager who takes responsibility for the overall completed project, so it is this person who must ensure that the combination of subprojects is viable and that they are properly co-ordinated.

Features of implementation

The implementation phase is the vital part of the project sequence. Some of its features are:

- **a big increase in activity, with more people and more machines**
- **a deeper organisation structure**
- **more formal management steps are required**
- **a big increase in rate of expenditure**
- **mistakes and delays are more costly**
- **bigger commitments are involved**

Emerging project and production infrastructure

At any particular time, a project will have reached a certain stage of development. The progress on a building will have reached a certain stage, an amount of software will have been developed, or some parts of an organisational change program will have been implemented. The project will have emerged to some extent. The emerging project is the project as it stands at a certain stage of development before its completion. It changes in form from day to day, from week to week.

Working directly on bringing the project into existence are people, machinery and many other elements. These elements are the production infrastructure, and some of them may form part of the project, for example, the materials used on the project. Other elements will not be part of the project but are a necessary part of bringing the project into existence, for example, plant that moves material around. In an organisational change project, little production infrastructure would form part of the final project, the changed organisation. Documentation is part of the infrastructure that will not form part of the finished project (except for the as-built and manuals, etc, which are part of the project).

The management of the activity of the production infrastructure is a key project management activity in the implementation phase.

So, when working on the project, there is the project itself taking shape. Acting on the project as it takes shape is the production infrastructure; it is the actions of this infrastructure that lead directly to the production of the project.

Project management is concerned to install the project infrastructure and to manage its activity to achieve the project. The purpose of the pre-implementation phase is to decide on and arrange for (though not necessarily actually set up) the production infrastructure to come into existence. Implementation will be concerned with the management of the activity of the production infrastructure.

Managing sequence and technology

It is not clear if knowledge beyond a fundamental grasp of the technology is always advantageous to a project manager—it can divert the project manager away from the job of managing. The important point is that the project managers can approach the pre-implementation phase with some confidence even if quite a lot of the technology is outside their experience.

The activity of the production infrastructure produces the project. If the production infrastructure is wisely chosen and working well, the project will be produced as expected. However, if the production infrastructure is unable to meet the project requirements, then no amount of inspection of the output at the end will produce a good project.

Measuring in a project sequence

There is a saying that if one cannot measure it one cannot control it. A wide range of measurements are possible. We can see different colours without being able to measure wavelengths of light, we can sense beauty without being able to measure it, and so on. Measurement is possible if choice is possible. It is possible to measure a wide range of things without having to put numbers to the measurement.

It should be noted that it is sometimes possible to go from a qualitative project requirement to a quantifiable (or measurable) requirement and then compare this with a measure of the capacity of the production infrastructure to perform. However, the definition of requirements is difficult and, in some cases, impossible. They are so difficult that they are rarely done explicitly, if at all. A project manager operating within an environment where this is not done as a matter of course cannot hope to introduce these definitions for all the work. In such cases, the project manager should choose a very small number of significant issues (as low as five) and carry out the exercise on them.

In a project sequence, measurement can be applied to either the emerging project or to the production infrastructure. Many times it will be necessary to rely on measurements of the activity of the production infrastructure rather than of the emerging project.

Emerging project measurements

There are two measurable situations here—one allows quite close control while the other is much more open to risk.

The most direct and reliable situation is where the progress of the emerging project can be measured and adjustments made to reach the desired outcome. This is often, though by no means always, the case in parts of engineering-type projects—work done to date on the emerging project is measured and then adjustments are made as required. Painting a surface and making adjustments while painting to achieve the desired outcome is an example. This is a situation where quality management can be easily applied and measurements of the emerging project can lead to high levels of reliability.

The other situation is where the emerging project cannot be measured, only the final project. Here measurements are taken at the end of the project sequence and a decision made based on those measurements, or it is decided by use. An election campaign is in this category—at the end, and only at the end, is it known whether or not the desired result has been achieved.

Production infrastructure measurements

Two situations are of interest here. One is where the activity of the production infrastructure is measurable and there is a *direct, known connection* between what the production infrastructure does and the emerging or final project. This is often the case in technical projects. An example might be building foundations; that they work is only known when they are loaded by the building (a little late to be relying on project measurements). This situation relies on measurements of the production infrastructure work to ensure reliability.

The other situation is when the activity of the production infrastructure is measurable but there is an indirect or only assumed connection between what the production infrastructure does and the emerging or final project. 'Indirect' means that there may not be a direct or causative relationship between what is being measured and controlled with the outcomes desired. The election campaign is an example here, as might be an organisational restructuring project.

Juries and measurement
Because the outcome may be so difficult to measure, so vague and difficult, subjective and personal, a jury may be established to decide the question. Juries are brought in to deal with the qualitative, and sometimes to gain agreement on the quantitative, measures. This is not often done on projects (except under the guise of the project control group, which actually functions in some circumstances as a jury). Project managers should give more consideration to juries. As an example, if computer manuals are meant to describe how to operate some software, a group of students of computer science could be the jury. It is important that the jury is both relevant and easily accessible. Unfortunately, this method has the difficulties associated with an inspection system—it does not help us very much in defining the sequence in measurable terms. It does not allow the use of a monitoring system to ensure that before the work is presented to the jury it will be acceptable.

Controllable aspects

The parts of the project sequence that are measurable and are in some way linked to the required outcome have to be identified—these parts are the controllable aspects. There are many projects where the emerging project is not directly measurable, such as organisational change, political campaigns and marketing efforts, and an indirect approach to measurement has to be adopted. Links, usually indirect, have to be set up or found between the project sequence and the required outcomes.

The relationship between the controllable aspects and the outcome may be imprecise or not understood. These relationships can often be found by means of Ishikawa diagrams, Work Breakdown Structures or some network methods.[61]

It is not easy to find controllable aspects. A two-step sequence is suggested that involves defining what is perfection and then putting that definition into operation.

1. Using perfection as a guide
It is everyone's dream that all will go well, that there will be perfect outcomes to the activity of the production infrastructure. But what does that mean? Knowing what is desired can be used to guide the project manager's actions. What would the management experience be if all of the production infrastructure activity went to plan? What would be the outcomes of a perfectly managed production infrastructure?

We are assuming that nothing goes wrong, that there is no deviation, no reason to take corrective action, which will help us decide what is being attempted and how to attempt it. It will help us to look at ways of managing so that things do go well, and also help in finding the relevant measures of satisfaction.

Starting off with perfection helps set up positive definitions of what is required. A positive framework is set up by identifying perfect outcomes where everything goes as planned. Perfection is later abandoned and the acceptable or required outcomes are defined. This involves defining the criteria for success in 'yes/no' terms, or in terms where there is a range within which the work is considered a success.

If a project implementation phase is to go as planned with all expectations met, some or all of the outcomes listed below must occur. In practice, and probably in theory as well, not all the outcomes can be met at the same time on the same project. Only parts can be fully satisfied, and sometimes not even that—at best, partial perfection. Trying to achieve the impossible is not unique to project management; it comes up in the quality management area as well, though it is rarely recognised. For instance, it is probably impossible to be totally innovative and totally reliable at the same time—the two qualities are essentially incompatible.

The following list is not exhaustive.

Before implementation

- The documentation is correct. This means that all the documentation fits together, that all the details and instructions it contains are compatible and technologically correct, and that those relying on it can do so with confidence. For example, the drawings of the piping in a chemical process plant show all the correct details and connections and the plant as described in the drawings would work. In a software project, the documentation would correctly describe the screens, the description of the data structures, the links between the databases and so on, and it would all be correct and work.

  The documentation would also be in a form that is suitable for the work to be carried out by the specialists working on the project. It should be easy to break up the documentation and present it to the specialists as they need it.

- Proper work methods have been selected and the correct skills have been identified.

- The subcontractors exist: the documentation can be given to a competent supplier who is willing to supply the services required and is able to arrange them.

- The subcontractor understands the documentation and the full extent of the work.

- Tenders were properly conducted: the process of attracting appropriate offers was legal, successful and an appropriate range of choice existed. The best subcontractors have been selected, and they have accepted the work.

- The project has adequate political support.

During implementation

- The subcontractor comes when required, does the work at the expected rate and finishes on or before the required date.

- The subcontractor does the work in budget. The price given for the work at the beginning is in budget and the work carried out is at that price.

- The subcontractor meets all the required project work standards. As the work proceeds, the subcontractor supplies the appropriate assurances and leaves people confident that the work is executed to the required standard. At the end of the work, the clients and stakeholders have a well-founded confidence in the work done.

- The subcontractor interfaces are correct. Where one subcontractor's work meets that of another, work conducted by others or work already there, the junctions are appropriate and proper.

- The various participants are paid the correct amounts at the correct time for the correct amount of work done. The money needed to make the payments is available as required.

- The as-built drawings and operating manuals are available as required by the project users.

- Off-site manufacture proceeds to plan. Co-ordination of their arrival, storage and installation proceeds as expected.

- No implementation changes are needed. The work as originally defined can be carried out without change; the work fits properly together.

- The wide range of materials required are available when they are required. They come in appropriate packaging, at a time that suits project progress, and are of the required standard. The project production infrastructure is able to receive the materials when delivered and is capable of storing them as required.

- Arrangements for staff to be available when and where required have been successful; both skilled and unskilled staff are available. The appropriate project introductions have been carried out successfully. Staff perform as required and leave in a positive frame of mind, with a sense of a job well done, when their services are no longer required.

- The challenges of the weather are met. The project production infrastructure is able to deal with the various types of weather experienced on the project.

- The public and various stakeholders are happy with the way the work on the project is carried out. There are good reports on the implementation phase.

- The various legal requirements are satisfied including health and safety laws, environmental regulations and taxes; privacy rules and industrial legislative requirements are satisfied.

- Political support is maintained. The key supporters continue to support the project and opposition has lessened. The project continues to satisfy the stakeholders' requirements.

Arising out of implementation

- The various risks and dangers have not materialised or where they did materialise they were adequately dealt with. The implementation phase had adequate insurance, adequate finance and adequate safety procedures.

- The project has enhanced industrial relations. Staff have been fairly treated, and feel fairly treated. Appropriate recognition has been given.

- Work was conducted in a safe manner. No-one has been injured as a result of work done during the implementation phase.

- Work was done as expected and everyone is satisfied. No legal actions or costly legal work arose and all issues are settled.

2. Operationalising the requirements

The discussion on perfection outlined in general factors to consider; it stated the ideal but it neither gave any information on how to reach it nor provided a measurable performance. It only helped focus on what is required. Perfection is limited in its ability to help manage the practical project; it is necessary to move from the perfect to the practical.

The depth of analysis required to identify a controllable aspect is easily recognised; when a specific action, person or organisation, or some other feature which is controllable has been identified (by means of, for example, the Ishikawa diagram), the search has been a success.[62]

There are three stages to this analysis. The first stage requires further precision about the meaning of the outcome. If, in developing the meaning of the outcome, a controllable element is identified, the analysis has probably gone deep enough. The second stage goes into issues that influence the outcome rather than simply elaborate it. In fact, the second stage analysis is often elaborating on the requirements of the project sequence. A third stage then actually nominates the required performance.

Three stages of analysis

The following is an example of how the search for controllable aspects might be conducted. The desired outcome is: As-built data and operating manuals are completed to the required standard. A definition is sought that will indicate when the desired outcome has been achieved; a sequence of activity and its accompanying measurement also need to be identified so that progress towards the outcome may be monitored.

- In the first stage of analysis, the meaning of the desired outcome is clarified as 'Ease of Reading'.

- Having identified the desirable feature, controllable elements need to be found—what leads to ease of reading. For example, the frequency of active verbs compared with the frequency of passive verbs, sentence lengths and the number of concepts in a paragraph are all aspects that influence ease of reading. These aspects are the results of the second-stage analysis and are capable of measurement—they are controllable aspects in the production sequence.

- In the third stage of analysis, requirements are attached to each aspect: for instance, nine out of ten verbs are to be active verbs, and each paragraph is to have only one concept . An acceptable or required level of performance of the parts of the sequence has been set. These levels can now be monitored and adjustments made as the writing of the manuals proceeds. Note that there is no logical, causal link between the measurements to be used in the sequence and the final outcome desired, ease of reading, but there is a possible, if not probable, link.

Practical control

Practical control requires to **either**

- *define the <u>required outcomes</u> (ie, the emerging project) in measurable terms of the activity of the production infrastructure (eg, the manuals example),* **or**
- *define the <u>required sequence</u> in measurable terms of the activity of the production infrastructure leading to those outcomes (eg, a cooking recipe),*

or **both.**

Political skills needed when measurement is unavailable
In situations where measurement is absent or not available, considerable political effort is required of the project manager to make sure the project or these aspects of them continue to receive support. The project manager should not be surprised that it is impossible to measure all outcome and sequence criteria and, in such cases, must engage in symbolic activity that is essentially a political and social process to ensure the existence of common understandings of the outcomes required and the sequence of getting there.

Select the critical few measurements
The project manager has to reduce the number of issues to a manageable number.

Assessing and deciding the production infrastructure

Assuming that the requirements have been examined and a project sequence and a production infrastructure have been selected, analysis is needed to determine whether they are adequate or whether adjustments are required. Such an assessment can be made using the following seven steps.

1. Identify the activity sequence to be examined.
2. Describe the sequence required, allocating responsibilities and filling in obvious detail.
3. Seek logical gaps and fill them.
4. Seek points that are consuming excess resources and reduce them.
5. Seek points of possible overload and resource them.
6. Balance the activities and resources.
7. Decide how to verify that the action and outcome have taken place.

In practice, the steps will not be worked through in the exact sequence outlined; one will jump around.

Step 1
Identifying the activity or sequence will often be fairly easy but there will be some boundary problems. If a sequence of activity cannot be identified, it is probably necessary to go back and develop the project sequence and the production infrastructure, or even go back to the planning and design stage.

Step 2
Describing the sequence required is quite difficult. It is not clear why, as the methods seem easy enough. However, experience in teaching the methods indicates hidden difficulties. The order of activities and

who is responsible for the activities are to be described; this can be done in a two-stage process. The first stage is to draw the logical sequence of activities in the precedence diagram form or in the critical path form. The second stage is to allocate the activities to the person or entity responsible. Now the action, the entity responsible and the time requirements have been identified.

Step 3

Much of the work involved in finding and filling gaps will emerge in the second step, drawing up the sequence. The logical gaps are the necessary activities which are missing, work which must be done before work can proceed, and so on. Examples might be the recognition that certain material must be available and no provision has been made to obtain it, that certain data should be available but no provision is there to get it, that certain people are to be involved but no provision is there to get them or they are elsewhere, and so on. These examples are fairly obvious once they are pointed out. However, that does not mean that these gaps are easily spotted or always filled.[63]

Two special gaps need attention. The first is easy to describe—the long lead item. The second is much more difficult to describe and identify—logical gaps.

On every project, there is the need to check on long lead times. This factor should be picked up in the planning and design phase and, in fact, the implementation phase may have to be structured around the long lead time items.

The second, logical gaps, are much more subtle. In many situations, if the ideas are not properly constructed or organised, if they have internal contradictions, things will not happen as expected. It is quite surprising how idea structures have such physical consequences.

One example is the need to have a consistent story or understanding in relation to the content of the project. Contradictions in expectations (ie, logical problems) will cause problems. In the implementation phase, idea structures might concern the authority to make payments and the authority to make changes. Other examples might be problems people encounter in introducing manufacturing methods, such as Just in Time. The simple idea needs many subtle shifts in attitude and approach before implementation can deliver the expected benefits.

Steps 4, 5 and 6

Having sorted out the logical part of the work, resources must be applied to the job. It will be most unlikely that the correct amount of resources will have been allocated. Resource levelling is an example of dealing with the undersupply of resources issue; oversupply of resources can be similarly dealt with. There is then the balancing work to be done. Where weaknesses are found, the proposed sequence must be strengthened. Where overkill or where great comfort is present, the sequence or its resourcing needs to be pared down. What is more likely to occur is that some parts are weak, some are okay and some are strong; however, the resources are finite. Resources will have to be shared around, so setting priorities and balancing work is required.

Step 7
The nature of the monitoring required must be established in order to verify that work has been done successfully. The project management approach is to monitor progress so as to be informed as early as possible of an impending problem so that corrective action can be taken. Not knowing, or finding out very late, drastically reduces the chances of fixing the problem. Forewarned is forearmed.

However, it is impossible to monitor unless a measure or objective against which to gauge progress exists. Hence the importance of establishing measures of acceptable or required levels of performance described above.

The work of the pre-implementation phase

During pre-implementation, arrangements are made for suppliers to supply, for people to be there when required, for working environments to be set up, for payment systems, for documentation systems, and so on.

The planning and design phase produced decision documents stating requirements in a form to which the supply industries could respond. However, they did not tell the supply industries how to go about making their response. A lot of work is required to transform the planning and design decisions into actions.

Getting pre-implementation under way
The planning and design documentation has to be examined and considered; then decisions must be made as to how to go about providing what is required. Essentially, arrangements for the project infrastructure are made and instructions prepared so that activity can start.

Project sequence
The work encompasses the following: identify the work to be done, the work packages, the tenderable set of work packages and the sequence of the work packages; take account of constraints; identify milestones and handovers, outcomes of various work packages and outcomes at certain milestones, and long lead items; and decide on budgets and allocate responsibilities.

Long lead items

It is necessary to make arrangements for their acquisition as soon as practicable.

Production site

Arrangements must be made for control of the site and for the supply of relevant services, such as telephone, fax, data lines, Internet connection, power, water, sewerage, copiers, computers and supplies (paper, film, etc).

Equipment

Identification must be made of the equipment and facilities to be provided and of the supplier; the hierarchy of use of equipment and access to equipment must be established; and arrangements must be made for the physical movement of materials.

Systems

The following systems are some of those that must be set up: an accounting system (procedures for the supply of funds to the project, identification of the payment system and procedures for the authorisation of payment, purchase orders, account numbers, etc); the documentation system must be established (orders, invoices, drawings, part numbering, instructions, variations, monitoring reports, procedures by which the documents come into being and by which their authenticity can be established, distribution requirements, etc); and storage systems, waste management systems, human resources systems (recruitment, terminations, various personnel documents, qualifications and medical certificates).

Implementation management structure

Arrangements must be made for the management of the implementation phase.

Main supplier

If not already done so, the main supplier must be appointed. (The main supplier must be in place before the end of pre-implementation; in fact, the main supplier has to make many of the arrangements listed here and thus would need to be appointed some time before the end of pre-implementation.)

Subcontractors

The following work must be done based on the tenderable set of work packages: identify the subcontractors required, extract the relevant instructions from the documentation, prepare tender documentation, call for expressions of interest, evaluate the responses, negotiate the final details, prepare contractual documents, obtain approvals to appoint, award the subcontract, arrange facilities so that the subcontractor has access to work, and arrange for the subcontractor to start work.

Procurement and expediting
It is necessary to identify how the project will obtain a wide range of items and chase up deliveries.

Production infrastructure activity
The initial activity of the production infrastructure must be planned.

Safety
The project safety policy and procedures need to be established, responsibilities for the management of safety allocated, and arrangements made whereby safety facilities meet or even exceed legal obligations.

Industrial relations
The industrial relations policy must be defined and developed.

Government
The necessary approvals and permissions must be obtained from relevant government departments (note the wide range of permissions and approvals required—access, noise, services, etc).

Training
Training needs must be identified and arrangements made for delivery.

Control system
Controls must be identified, as must monitoring and reporting procedures; methods to verify progress must be established; and required skills and documentation identified.

Project manuals
On larger projects, manuals are often prepared to set out the procedures for various systems. These manuals often cover issues such as the funding process, payment sequence and change sequence. But, as stated earlier, they should be limited in size and detail.

Dual organisation set up
The client and supplier organisations must first be identified and then set up.

Approvals
The required client approvals must be obtained.

Some of these items require different work from each of the client and the main supplier. The appointment of the main supplier is the client's responsibility. How this is done depends on the contractual and procurement strategy employed; this would be decided either in the feasibility phase or in the planning and design phase.

Handling some of the items

As can be seen from the above list, much has to be arranged and usually arranged quite quickly. The following are comments on some of the items.

Work packages

A work package is an identifiable, measurable and controllable piece of work which can be integrated with other pieces of work in the project sequence. Clearly, there can be bigger and smaller work packages and also work packages which can be further broken down into smaller ones. Industry-specific knowledge is needed in order to identify a work package. Typically, a work package will involve like skills (that is, related to a specific technology) or will involve the supply of a complete function.

In renovating a kitchen (see figure 4 overleaf), all the work could be in one package. Alternatively, the work could be divided up into a set of work packages based on the trades used to effect the renovation.

A work package can be tendered singly, or a set of work packages can be combined into one tender package. A tender package is an amount of work which can be let out to tender; it should not cut across a work package, rather, it should be made up of a set of whole work packages.

From the client project manager's point of view, the more the work is broken down and let as small tender packages, the more integration and co-ordination work to be done. As the tender packages get bigger, the responsibility for co-ordination passes to the supplier and the project manager has less control over the detail of the work.

The choice of tender package is very much dependent on industry practice. In the first place, it depends on industry's ability to handle the work package. Secondly, there is a cost penalty associated with larger packages, in that combining work packages leads to the tenderer levying a management charge. Sometimes there is synergy in combining work packages, but this is difficult to recognise. A tender package is also a means of allocating risk—where work packages are combined, some of the integration risk is transferred to the tenderer.

The choice of tender package should also take account of scheduling requirements; flexibility should also be part of the tender requirement. The project manager needs to consider safety when entering into contractual relationships.

One of the key jobs in the pre-implementation phase is to identify the sequence of work and arrange for the project infrastructure to be put in place. Much of this work will be achieved through work packages.

FIGURE 4
Transformation from functional elements to work packages

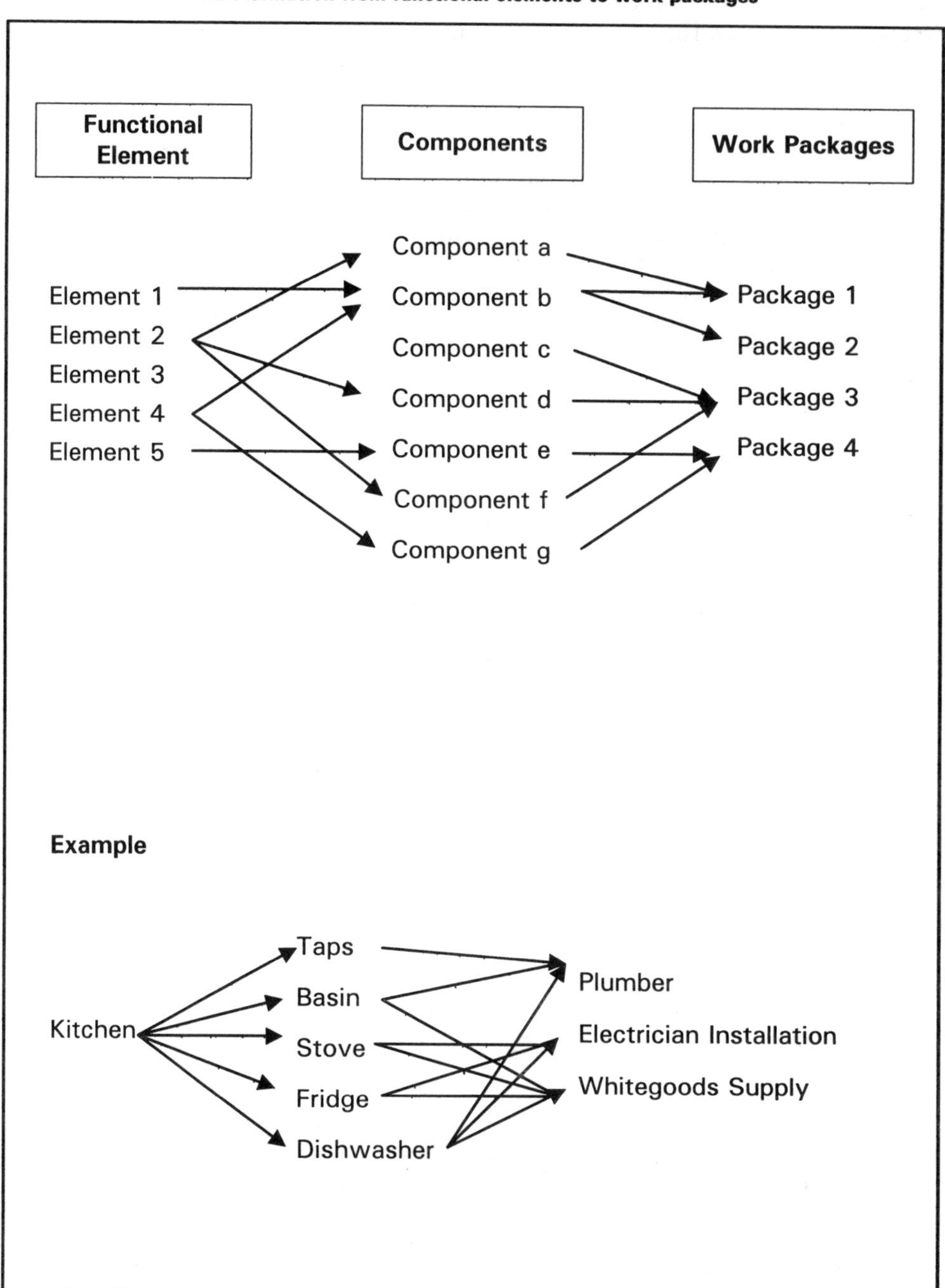

Systems

Systems cover a wide range of business practices, many of which are standard in organisations going about their everyday business. Those involved in projects have to set them up from scratch for each project unless they have developed portable systems or they are able to connect into the parent organisation's systems.

Many of these systems are generic and independent of the project under consideration. Certainly, in accounting systems and document tracing systems there is considerable commonality from project to project. So, from the project manager's point of view, there is considerable merit in involving oneself in a continuous improvement process in relation to developing these generic information systems.

While there is an advantage in using the parent organisation's system, it does have the drawback that some control over the system is lost.

What needs to be emphasised, however, is that these systems are the brain of the project management work. They provide the vital data on which decisions are based.

Long lead items

There are two types of problem here. Firstly, there is the problem of acquiring an item which takes quite a long time to make and deliver, and secondly, there is the problem that, while there is nothing to be obtained, there is a long lead time before work can start.

An example of a long lead time in relation to access might be a project on the railways. Obviously, work cannot start at any time on a particular piece of track. Joining one piece of track to another on a railway line requires closure of the line, which must be organised in advance. Many railway companies have a 'possessions' committee whose job it is to decide which track will be closed, when it will be closed and for how long. It is then decided who has possession of the track.

The project manager must recognise the existence of long lead items in relation to access, and that they vary from industry to industry.

Output: project implementation manual

It probably helps to maintain a sense of control over what is going on in the pre-implementation phase if a set of documentation is prepared, a kind of project implementation manual. This manual could be a summary of the project to date and act as a master control document for the implementation phase. It could record the various decisions, and contain a range of source documents. Obviously the manual need not be seen as a book in the traditional form, or even confined to paper in loose-leaf folders. It could consist of a range of documents in many forms—

paper, electronic, film or models. The project implementation manual is essentially evidence of what has been achieved.

There will probably be two such manuals: one prepared and maintained by the client project manager and the other prepared and maintained by the main supplier's project manager. The manuals will be a growing database which will need careful management. They will be a very useful information tool, backing up the project manager's control.

The manuals may include, among other things, some or all of the following:

- evidence of commitments
- names of key personnel
- financial arrangements
- project operations manual
- interface schedules
- authorisation processes
- documentation structure and flow
- mobilisation plans
- responsibility allocations
- equipment and material arrangements
- networks allocated between participants
- resource levels
- control budget, control codes and control account numbers
- cash flow projections
- time schedule
- change procedures
- contracts as signed
- schedule of standards that apply
- special standards
- safety policies and procedures
- industrial relations responsibilities
- schedule of schedules

Dual organisations

The client has to separate itself from the suppliers, a step that must be achieved before implementation begins. The implementation phase requires two organisations be set up; these can be quite large and complex but are always temporary. These project organisations cut into other organisations so it is difficult to say exactly where their boundaries are.

It is, of course, technically possible for the client and the suppliers to remain as one entity. However, this is not seen as a good model: there are quite different roles to be performed and it is advisable that these roles be separate. One organisation looks to ensure that the client's interest is expressed and followed through on, and the other is concerned to deliver the project and profit from delivering that project.

On many projects, this separation will have already occurred—it really depends on when the main supplier was selected. In contracts dealing with design and supply, the contractual relationship would have been set up quite early and separation would also have occurred at a very early stage; there would be a clear client organisation and a clear supplier organisation. The issue becomes clouded when the planning and design consultants are separate from the main supplier. In this situation, the main supplier is chosen quite late, often during pre-implementation. The main supplier is then clearly seen as separate from the client. However, the planners and designers often see themselves as part of the client organisation and not as suppliers. Legally, they are suppliers, yet are professionally supposed to act on behalf of the client. So some overlapping in roles will occur.

This model of two organisations, each dancing with the other, managing the project, is fairly obvious when the project is being 'bought' by one organisation and supplied by another. Purchases of buildings, of hardware and software systems, and of large capital goods usually fall neatly into this category. There is clearly a customer and there is clearly a seller; the customer is the client and the seller the supplier. The customer is always right as long as the customer is prepared to pay. The customer sets up an organisation to act as its representative in the project acquisition process just as the supplier has a project organisation to act as its representative and to deliver the project.

On internal projects, the need to have the two organisations, the need to make this separation, is sometimes not understood. But once in the area of projects, the logic of the project management principle that the client be separate from the supplier becomes evident. When it reaches a point where an organisation both pays for a project and supplies itself with the project, it is necessary for it to bifurcate, for one part to become the client and the other, the supplier. The two-organisation arrangement is necessary essentially to ensure accountability and to provide discipline to the project acquisition process.

To achieve the split, a 'client' part of the organisation is set up. This part is responsible for the delivery of a service to the external world. For example, in a water supply organisation, the client part might be set up as the entity responsible to supply water to a district. To do this, it runs the production facilities and gets paid for the delivery of the water; it also requires that a project be undertaken to upgrade and improve the production facilities. In this example, the client part is expected to deliver a service efficiently and to acquire a project efficiently. Acquiring the wrong project, or acquiring the project at the wrong price, both lead to danger for the client part. On the other hand, the supplier of the project is kept in check by the threat of the client part going to another supplier.

The main supplier

Usually, the work is undertaken by the primary contractor who employs various subcontractors to execute the work.

The role of supplier has been typically undertaken by organisations on a fixed-price basis or on some other, but essentially 'buyer beware', basis. There are now organisations that perform this as a management role and set themselves up as professional consultants.

The main supplier organisation on capital projects has to mobilise a great number of people, administer the interfaces between subcontractors, make the payments to subcontractors, claim and obtain payment from the client, and supply materials and locations for work. On organisational change projects, the supplier, although probably quite small, needs to mobilise the skilled people needed, make upfront payments to them, and make claims from the client.

The main supplier must recruit the relevant skilled people for its own organisation. It must provide them with appropriate conditions and possibly accommodation. Hence the supplier organisation has a personnel function, a function that becomes more elaborate when one is managing overseas projects.

Size of supplier organisation

The supplier organisation (main supplier plus subcontractors) can be very large. Its full extent is hidden because of the subcontract structure. It has a diverse range of skills ranging over technology and administration, with the added complication that the organisation is temporary.

The main supplier project manager has to build up this organisation, hold it at a peak size for the required length of time and then disband it.

Where possible, the work of the main supplier should be managed by a team or small group structure. The best situation is where the whole team can be allocated exclusively to the project. In this case, the project manager can probably get a project office with the team located in this one office. This allows a strong team spirit to develop and strengthens commitment to the project.

However, some personnel have to be shared between projects. These people are often those who possess specialised skills; the project manager must negotiate carefully for their services. Many of these people then end up having more than one boss, with several people putting pressure on them to meet their own requirements. In cases like this, the project manager should not leave it to the people themselves to negotiate—that is the job of the project manager.

As the project gets larger, divisions are set up within the project structure. Often the technical people split off from the administrative people, and specialist skills are grouped together. There are line and staff personnel. This is not the only way these things can or should happen, but it is often the case. An alternative is for the project manager to see the overall project as consisting of subprojects, and then dividing the project into subproject groups. The purpose here is to maintain a mix of skills in each project team, where the focus will be on the project rather than on the discipline area of the specialist.

Subcontractors

The main supplier organisation has to bring all the elements of the technology of the project together and take it to a point where it will work (sometimes the client organisation will bear the responsibility that the project as assembled will work). This is the major part of the supplier's work. This means that the main supplier must know how the particular technical areas work, how information is managed in those areas, and know the key aspects to control. On projects, this is all usually taken on and delivered through subcontractors.

Having subcontractors is not essential to project management processes; however, subcontracting has grown and is now widespread. The supplier organisation could elect to supply the whole project through its permanent staff. While this has advantages in the administration of some quality aspects, it exposes the supplier to serious risks because of fluctuating demand. Essentially, it is because of fluctuating demand that suppliers have turned to subcontracting.

The supplier has to acquire the services of subcontractors and make arrangements for their accommodation within the project. This accommodation involves providing facilities for both people and machines, and providing information. In some industries, the skills of the subcontractors are low and the main supplier has to provide them with considerable support services.

The supplier also needs to establish a system to deal with the various interactions of the subcontractors. These will be managed by much one-to-one talking during the project, and also in project meetings. What is interesting for people working as part of the main supplier organisation is that they will always be the focus of technical queries about the job. These queries start at the beginning of the job and continue unabated until the end. It is extraordinary the number of details that need attention.

The supplier personnel are inevitably, at some stage, in a position where they have to accept or reject the work of the subcontractor. Where the supplier is providing the guarantee, there is pressure on its personnel to be competent in their acceptance and rejection of work. Where it is simply a matter of whether or not the client will accept the work, the supplier might call on the client to accept the work as it progresses (wise client personnel will tread warily here).

Managing the work

Much of the work of the pre-implementation phase is management work. It involves planning and decision-making. If there is no main supplier, the responsibility to initiate the pre-implementation phase rests with the client project manager, the main responsibility being to find a main supplier. If there already is a main supplier, the responsibility to initiate will depend on the contractual arrangements; it would be in the client's interests to retain a right to approve what is arranged in the pre-implementation and implementation phases.

Assuming that the main supplier is in place, the client project manager and the main supplier project manager go about organising themselves for the implementation phase. To do this, they have to carry out the work of the pre-implementation phase. On small projects, much of this work can be done by one or two people, but on larger projects, teams will have to be set up to get things organised. The main supplier is more likely to need a team than the client.

If a team is set up, the project manager must identify the tasks, quality of work and skills, make selections and appoint people; the tasks will also have to be allocated.

Maintaining an overview

The project managers must maintain an overview, although it will be difficult due to the fact that the detail of the project is suddenly exploding and demanding attention. Cost and time especially must be kept in focus, and risks and capability regularly assessed.

During pre-implementation, the participants are confronted with having to again make trade-offs. As pre-implementation proceeds, they discover conflicting requirements, impossible expectations and so on. Typically, they find time expectations particularly pressing because stakeholders see themselves as having made decisions and want the project—now!

Plans and arrangements must have some flexibility, otherwise things are vulnerable. It will not be possible to introduce sufficient flexibility to feel comfortable, so the project managers should identify the risk areas, put them into some order of priority and then make a great effort to ensure they are well-managed. Finally, approval to proceed should be obtained from the client.

Chapter Seven

7

Implementation Phase

Now the rate of activity hots up. The rate of expenditure increases, more and more people become involved and there is work to be done everywhere. The project presents interesting physical and social problems, all demanding the urgent attention of project managers. Project managers need to take an overall view and recognise that other issues are there to be dealt with, not just the internal ones demanding immediate attention. A major issue is the continuing external support for the project.

There is never enough time during implementation, a fact that all project managers never lose sight of. Knowing it, they make the next step, which is to set priorities. During implementation, they will be constantly reviewing the priorities, always facilitating the forward movement of the project sequence.

On most projects, the purpose of this phase is to carry out the designated work, work described in specifications, drawings, manuals and coding plans. Where what is to be done is quite clear and known methods exist for its execution, the main work is arranging for the work to start, keeping it going, monitoring and controlling its progress, making payments, and assuring that the work is up to the required standard. It involves dealing with subcontractors, some good and some not so good. Much of this work is based on an understanding of the governing contract and administering the job to satisfy the contractual requirements. It involves dealing with some changes, it involves arguments about the meaning of contractual conditions, it involves moving funds around the budget and managing the contingency, it involves time charts of various forms and many, many meetings. It also involves some degree of concern with efficiency (but little with overall project effectiveness—that has been decided). It involves a lot of talking and living with the feeling that the paperwork will never be up-to-date. It is utterly engaging for some.

On projects where the main work is not quite clearly laid out, where the work is only described by some general conditions to be achieved, or where what one does depends upon what one finds, there is a need to be able to react quickly and change direction if necessary. Usually on these projects, a way of solving the problem is formulated or a way of going about the work. This direction is followed until an alternative emerges and takes over. Effectiveness here is a key issue. The activity of searching for a lost boat at sea will change in the light of what is or is not found. Moving an organisational change along might encounter significant unexpected and powerful resistance which indicates a need for reconsideration, a shift in approach. This type of implementation phase is much more difficult to control in terms of knowing the cost framework, the time framework, etc. Despite this, there will be a need to keep things going in the direction decided on and, probably just as important, a need to know the route travelled so far. In addition to constantly reviewing effectiveness, arrangements must be made for the work to start, to keep it going, to monitor and control its progress, make payments, assure standards, deal with subcontractors, move funds around the budget and manage the contingency. It will certainly involve dealing with proposed changes and many, many meetings. All this is also utterly engaging for some.

As might be expected, an implementation phase may contain both elements—it can contain highly defined work as well as work which needs to be clarified.

The project manager, despite having done excellent planning, is to some extent pushed into a reactive mode. In the implementation phase, peace and stability are not on the agenda for the project management team. There are usually long hours of work, and home life suffers.

The work of implementing the project

The implementation phase is conducted by two parties: the main work is done by the suppliers with complementary, but much less voluminous, work done by the client. So in examining the implementation phase, the work of both the suppliers and the client and the interaction between them need to be examined. Each side is represented by a project manager, the supplier project manager and the client project manager.

The work is done by the two organisations whose work often interconnects, such as in approving payments to be made, accepting work, dealing with changes and agreeing extensions of time. A project is a joint effort, not a solo performance.

Only the key focus of the work is mentioned here. For it to actually happen requires the organisation of what is described here plus the organisation of many other supporting activities, both internal and external. The work described below is separated along the lines of the work done by the two organisations, client and supplier.

Implementing the project: the client

The following outlines some but not all of the client functions.

The project is proceeding as approved

The client project manager's main function is to be satisfied that the work is proceeding as expected, that progress is adequate and that the work is of the required standard, and to assure the client accordingly.

The client project manager has to take corrective action when problems arise or when problems are expected. This will involve inspections, monitoring time progress, monitoring costs, checking that all the project is being installed, and that any observations required take place when required. The project manager must also have access to and an understanding of the project contractual documentation.

Payments

The client organisation has to make payments, usually on a progressive basis, and usually to the main supplier (there are some situations where the client will make payments directly to subsuppliers).

Project elaboration

The client organisation needs to be able to further elaborate its requirements—a very important function of the client project manager, and is often quite difficult. The organisation must be able to answer technical questions; examples range from how the acquired project will be serviced and maintained, to standards (in the electronic consumer market, this is a very difficult area and considerable resources are devoted to reaching agreement on the standard), to the industrial implications of introducing some work processes, to safety implications and rules of operation. The client organisation clearly needs access to skills in the technologies of the project so that it can respond from firm ground.

Variations

The client organisation has to deal with the need for changes and also with the need to reject unnecessary changes. This involves agreeing the nature of the change and the pricing for the work of the variations.

Implementing the project: the main supplier

The following outlines some but not all of the main supplier functions.

The work of the work packages

The work of the work packages is the main stuff of the project. It is the technical component and is essentially what is bought when the project is bought. To get work done on a work package, the appropriate skills and materials for the execution of the work must be organised, people found to do the work (usually subcontractors), materials supplied, sufficient access to do the work provided, services and equipment provided, the work supervised so that the required quality standards are attained, and close-off achieved when the work is finished.

It is also necessary to interface the work of different work packages, to integrate off-site manufacture with on-site work, to make sure that the work of one subcontractor fits in with that of another, and to expedite the delivery of items.

The main supplier has to pay the subcontractors and other employees, and pay for some of the supplies; this requires having sufficient funds available. The supplier has to lodge claims for payment with the client.

Executing the work packages is a very busy activity.

Approval of variations
Because projects are evolving, project management often involves dealing with changes. These can arise from a wide range of sources and the supplier needs to get them formally agreed to and paid for. There are also contractual reasons, such as liquidated damages, why a supplier might apply for variations, particularly to extend the agreed contract time.

The variation application has to be prepared and justified. Estimates of costs in terms of both money and time have to made, and prices established and agreed.

Quality product
The work has to be delivered within the agreed standard, otherwise the supplier will be faced with a range of complaints which will have to be dealt with.

Doing the work of the implementation phase

To actually do the work of the phase, of course, more work than that described above needs to be managed and carried out. Activities are carried out with effects that do not end up as part of the project. An example might be employee morale, a key concern on any project; however, morale does not end up in the project. The activities listed above are directly associated with the project. The project manager manages these as well as other activities that are necessary to maintain the conditions of production. Those activities directly associated with the objectives, plus other supporting and necessary functions are mentioned here, and their management discussed in more detail in the next section.

The work is divided into two parts: internal work and external work. Internal work is focused on getting the project work done while external work ensures that the project continues to be supported from the environment around it.

Remember that the ideal scenario is one where the documentation is correct and the subcontractors understand it, the project has adequate political support, the subcontractor carries out the work on time and in budget, the subcontractor carries out the work to the expected quality, subcontractor interfaces are all correct, payments are all made appropriately, as-built data and operating manuals are completed to the required standard, off-site manufacture proceeds to plan, no implementation changes are needed, material is available when required, people are available when wanted, the challenges of the weather are met, the public supports the work of the implementation phase, the law is observed, political support is maintained, no risks arise, industrial relations proceed smoothly, work is successfully conducted in a safe manner, no legal actions arise and all is settled—and so on. But it is never this way. Sometimes things are well planned and sometimes they are not. Either way, a lot of work is required on both sides.

Performing internal functions: the client

The client internal functions are exercised through the client project manager. They include:

- ensuring that the project is proceeding according to the approved plan;
- financing the project sequence;
- making payments;
- elaborating the project;
- maintaining clarity;
- maintaining enthusiasm and morale;
- deciding on variations;
- setting up and running a client accounting system;
- setting up and running a client documentation system; and
- setting up and running the client temporary organisation.

Performing internal functions: the main supplier

The main supplier internal functions are exercised through the main supplier project manager. They include:

- carrying out the work of the work packages;
- maintaining project momentum;
- organising the translation of contractual documentation;
- managing subcontractors;
- making payments;
- liaising;
- doing work on time;

- administering the contract;
- doing the work in budget;
- ensuring that the subcontractors continue to supply;
- setting up and running the materials handling system;
- setting up and running temporary accommodation and other staff services;
- running personnel systems;
- ensuring industrial safety;
- managing industrial relations as they affect the project sequence;
- setting up and running a temporary organisation;
- obtaining approval for variations;
- organising the delivery of a quality product;
- financing the project sequence;
- setting up and running the accounting system;
- setting up and running the documentation system;
- periodically issuing project control statements; and
- running the project management information system.

Performing external functions: the client

The client external functions are exercised through the client project manager. They include:

- maintaining clarity and enthusiasm;
- ensuring clear ownership is maintained;
- maintaining an acceptable public profile for the project;
- ensuring the continued supply of resources; and
- ensuring continued support for the client project manager personally.

Performing external functions: the main supplier

The main supplier external functions are exercised through the main supplier project manager. They include:

- ensuring the market can supply;
- preparing the receivers of the project; and
- ensuring client acceptance.

Managing the work

All the above has to be managed, and managed between the two organisations. In any one industry, many of the tasks to be managed are standard, repeated from one project to another, and many cross industry boundaries. On a specific project, there will be other issues to manage. It is therefore worthwhile for project managers to develop their skills in the generic areas as experience grows; this allows more time to be available to deal with the specifics from industry to industry.

Getting the project sequence going

The main supplier project manager has responsibility for most of the work initiating the implementation phase. This project manager may be prodded into action by the client project manager. Contractual conditions usually lay down deadlines by which the implementation phase of the project sequence work must start, and it is the client project manager's responsibility to ensure that the work does in fact start. If it does not, then the client project manager may have to take contractual action to initiate it.

Something the client project manager can do to help things along is to organise a project start-up meeting. This allows the main people involved to get to know each other and to negotiate in more detail how they will work together. The main supplier project manager should also have a separate project start-up meeting which involves the subcontractors.

Getting the project up and running involves putting all the temporary facilities in place and bringing the initial people on board.

Managing internal functions: the client

The client internal functions are managed by the client project manager.

Project is going according to plan

During implementation, most of the effort of the client organisation is concentrated on the control cycle, examining the quality of the work produced, the timing of work and the extent of the work done. When all is going well, this activity is easy enough; it becomes difficult and sometimes heated when things are not going to plan.

The client project manager must identify the plan, measure progress against that plan, form a judgement as to how things are progressing, and then, if necessary, take corrective action. This manager, though, is not the main supplier and therefore has to work in an indirect, rather than a direct, way to achieve the desired outcome. This is a real problem and dilemma for the client project manager.

The client project manager is responsible for the outcomes of the project sequence and for the choice of overall project sequence (as against its detail developed by a subcontractor), and can act in quite global ways on the sequence by changing some components of the project infrastructure. The client project manager can set up additional parts, remove parts and so on. However, this must be done carefully within a quite tightly defined legal structure and system.

However, having put the project infrastructure in place, the client project manager can then only act indirectly on the internal workings of the project infrastructure, at least, those parts that are under the control of the main supplier. This is a very important division of responsibility and the client organisation representatives need to be trained so that in their interactions with the supplier, and with the subcontractors, they clearly maintain that the responsibility for the detail project sequence always lies with the supplier. The client project manager, having agreed to pay for the responsibility to be taken by the main supplier, should not take it back by reason of default or by moving into the supplier's area of control.

This problem has to be overcome and the client project manager has to get in close to the detail project sequence and influence it without taking on responsibility for it. It is too late to be rejecting substandard work at the end of the sequence—time is always a pressure. Any substandard work has to be found while the detail project sequence is taking place.

The client project manager must therefore put in place the following sequences of activity and ensure that they are properly resourced.

- Understand the project documentation.
- Monitor progress.
- Accept or reject the work: accepting or rejecting the work of the supplier is a client organisation function. This organisation needs to inspect and make technical judgements on the quality of the work. The process of acceptance requires technical skill. While much of the formal acceptance takes place in the next phase, in practice much of the acceptance procedures take place during implementation.
- Intervene if needed: indication that intervention is needed may come from observations of the emerging project outcomes or from the behaviour of the production infrastructure. Thus the monitoring may be a basis to push for and negotiate a change to the production infrastructure. Where progress is not satisfactory, the client organisation must apply pressure on the supplier organisation to conform (usually with the ultimate threat to withhold payment, a very blunt instrument), without itself taking on the supplier's contractual obligations. This process of applying pressure is usually quite social in nature and relies to some extent on negotiation.

Finance is available

The client has to make sure that the cash is available to make payments as and when required. Quite considerable sums of money will be involved, varying in magnitude from month to month.

A cash-flow projection is required to give a forward estimate of the cash demands of the project. The actual demands will vary, so at least two projections should be looked at, the fast expenditure and the slow expenditure cash-flow models. The problem the client project manager faces is that, while the order of magnitude is known, the exact requirement is unknown in advance. The project manager has to arrange banking services to provide the cash as required; banks charge fees to provide this variable cash demand facility.

One of the defences against being surprised by the demand for payment is to have a delay mechanism built into the payment process. While wages and salaries have to be met on demand, the subcontractors' bills can be met with an in-built delay. In effect, the subcontractors advance credit to the project. It is normal, but can be negotiated otherwise, that work executed in one month is billed in the next month and then paid in the following month, thus about a six-week delay occurs.

Having negotiated the variable cash demand facility, two things can go wrong. The first is that the rate of expenditure moves outside the range expected, and the second is that the overall cost blows out. Each of these situations demands urgent negotiation with the financiers. What is important is to avoid surprises—it must all appear to be under control!

Management of payments

When the work is done, it must be valued so that the appropriate payment can be calculated. This draws on the resources of technically capable staff who then make a recommendation for payment, or might approve payment. In each industry, there are specialists who provide valuations of work in progress.

Given that bills are turning up quite a lot later than the completion of the billed work, the valuations during the main body of the work do not have to be very accurate. By the time the bill arrives and is due for payment, the client's debt to the subcontractor has increased substantially beyond that being billed. If the work is technically satisfactory, then agreeing payment should not be difficult during the mid-phases of the subcontractors' work. However, considerable care must be taken with the last few payments. Usually there are retentions on some minor outstanding matters; payments need to be retained as motivation for completion.

The main supplier does not always provide credit to the project as a result of the delayed payments process. Various devices can be set up to deal with this. For example, where the main supply organisation is an organisation which supplies on a professional basis, ie, it is paid by fee, various arrangements can be made for subcontractors and other suppliers to be paid. One such arrangement might be the use of a trust fund which the organisation draws on to pay its own ongoing project expenses and the subcontractors. The trust fund is topped up once a month by the client on the presentation of approved invoices and other bills.

Where the client is paying on the understanding that the payments are to be forwarded to subcontractors, the client project manager may have to regularly require proof of such payment. This could be in the form of a copy of a receipt or, better still, a certificate of receipt of payment issued by the subcontractor.

Payment of accounts requires continuous ongoing activity during the project sequence and usually requires separate personnel.

Project elaboration and clarity

While stating exactly what you want sounds simple enough, in practice there are many difficulties. Queries arise about the work of the specialist designers and planners, who therefore need to be available during implementation in order to answer them.

A bizarre example of the continuing need for the designer's involvement concerns the use of CAD drawings. While the CAD system itself may 'know' all the dimensions, they may not be put on the drawing. Thus, although the drawings are drawn to scale and are correct, vital data is missing. It is necessary to go back to the CAD system to find the dimensions (assuming of course that the database still exists).

Most project contracts have time limitations on how long the supplier must wait for an answer to a query. There must therefore be sufficient resources to answer the question both technically and in time. Failure to properly manage project elaboration will lead to confusion, delays and extra costs.

Enthusiasm and morale

Maintaining enthusiasm involves being aware of what others want out of the project, holding out the possibility of that want being satisfied and hopefully satisfying it. Keeping enthusiasm alive involves aiming at a long-term objective and identifying the sequence of short-term goals to be achieved along the way. It also involves giving people news, both good and bad, on the progress of the project. When there is bad news, people need to be treated sensitively, and taken through a process whereby they can accept the set-back and get on with the job. Keeping up enthusiasm and morale involves making sure that the various project stories are compatible with one another; any contradictions that crop up must be quickly sorted out and clarified. There is also the need to make sound, defensible, but maybe unpopular, decisions as and when they are required.

On longer term projects with, say, an implementation phase lasting over six months, there is a need to be quite explicit about the management of enthusiasm and morale.

Decisions on variations
Changes happen on projects, quite unavoidably, that must be managed properly. Any proposed change should be examined to see if it is really required: if so, it must be accepted by the client with agreement on cost and extra time requirements; if it is not required, then it is rejected. The process of acceptance and rejection of changes is often called 'variation management'.

Speed in processing variations during implementation is important because of the rate at which resources are expended daily. Managing variations is a central project management function involving both the client project manager and the main supplier project manager. A variation may start in the client organisation or in the main supplier organisation. The variation is then formally proposed (ie, documented properly), which should involve it in having a number attached to it. The client should maintain a variation schedule.

A recommendation is then prepared by the client project manager for action by the client. The location of the authority to agree to the variation depends on how the client organisation is organised. A general principle is that the variation should be decided on as close to the project as possible, preferably by the client project manager.

Difficulty can arise when the client approval/rejection process is much slower than the rate at which variations are generated. In preparing a variation recommendation, it is useful to point out its implications, but it is practically impossible to specify the full implications when the variation management is slow. Slow variation management also leads to considerable confusion if answers are not provided. On a fast-track project, this can be critical.

While waiting for a variation to be considered, a fairly general rule for the client project manager is to demand that work, as approved to date, continue until a change is approved. No work is to be stopped because something might change.

Having a set procedure is useful to the client project manager in that it can be used as a lever against other executives introducing changes without having them properly assessed.

Managing variations well is one of the hallmarks of a good client project manager.

Client accounting system
The client project manager obviously needs to set up and run an accounting system to provide the data as required and to allow proper recording of the financial obligations entered into as the project sequence proceeds.

Client documentation system

Documentation systems are quite difficult to design, at least design so that they motivate people to continue to follow the rules of the documentation. This system has to deal with, among other things, orders, invoices, drawings, part numbering, instructions, variations, monitoring reports, procedures by which the documents come into being and by which their authenticity can be recognised, distribution requirements, equipment inventory and so on.

Client temporary organisation

Probably the most important thing here is to recognise the need for this temporary organisation. The client project manager has to negotiate the resources for it.

Managing internal functions: the main supplier

The main supplier internal functions are managed by the main supplier project manager.Some of the more general matters to be managed are maintaining project momentum, subcontractors, cost, time, quality, risk, documentation, industrial relations, variation/alteration management, team management, materials handling, industrial safety, contract administration, production and productivity and handing on. On a specific project, there will be other matters.

Oversight of the work of the work packages

This is the main driving force of the project. Assuming that the external environment remains favourable, the success of the project is determined here. The main volume of the project work is done through the execution of the work packages, and their management becomes the management of subcontractors plus the management of the interfaces between subcontractors.

The subcontractors who are doing the actual work of the work packages have to mobilise, find the relevant skilled people, usually supply the required material, translate specialist documentation into instructions which the skilled and unskilled can understand, direct the work of the skilled and unskilled, expedite work, anticipate problems, respond to problems, redesign work to suit available skills, pay the people working for them, make claims for payment, get agreement from the main supplier to accept the work done, make good any defective work, tidy up and leave.

Oddly enough, one of the main threats to the subcontractors is boom conditions—at these times, staff are unavailable and costs of both labour and materials escalate.

For every subcontractor, the main supplier project manager has to receive the subcontractor into the work-place, provide access to work, provide the general project sequence infrastructure, measure progress and take corrective action, measure productivity, seek opportunities for improvement, control the work of the subcontractor in terms of time, cost and quality by inspecting and monitoring the work, execute the contract requirements, expedite and accept or reject the work of the subcontractor, issue change orders, make payments, get work made good and make the final payment.

In managing the subcontractors, the main supplier project manager has, on one hand, to stay at arm's length and, on the other, be quite closely involved.

In addition to this, the supplier project manager has to make the work of the subcontractors fit together. This can mean marrying physical details, providing consistent explanations on social projects, ensuring data structures are compatible and in many other ways making sure all works well together. Interfacing involves having staff identifying interfaces (probably using responsibility matrices), liaising across interfaces—even task forces dealing with particular interfaces—checking that all is well and taking action if it is not.

Access by the subcontractors to production infrastructure can be a key issue in interface management. Multi-activity charts, which show who is using a facility and when they are doing so, are helpful. Physical clashes are unfortunate and always give the opportunity to criticise management; they are easy to avoid when there are only a few interfaces. When there are dozens of interfaces and dozens of claims on infrastructure, clashes easily occur. Good management is needed to avoid them.

Another area of work to be co-ordinated is off-site work, which occurs on practically all projects—long lead items are usually off-site work.

The main quality strategies should be implemented within the work of the work packages. Where repetition is present, and in many projects it will be present at the operator level, there is the opportunity to apply standard TQM techniques. Where the work of the project sequence is always one-off, there is considerable advantage in the idea of the downstream customer and in that customer having a say in what is expected. Personalising this contact across interfaces provides a very useful motivator.

Subcontractor time performance

The work packages are co-ordinated by quite detailed time programming. While the client project manager may be happy with programs which are monthly based, the main supplier project manager has to work with, at the longest, weekly programs which may be presented as rolling fortnightly programs.

Programs help to identify peak demands on resources, particularly personnel. Resource levelling is commonly carried out to try and ensure that the peaks do not exceed supply.

Another time-management strategy is called 'crashing tasks'. This is needed if a 'work around strategy', a 'simplification of work strategy' or a 'low-level overtime' strategy will not suffice. In crashing tasks, efforts are made to catch up on lost time or to achieve a specific task very quickly by putting on a lot more people (or other key resource, such as machinery or computers). The hope is that the work will get done more quickly by having more people on one task or by starting a wider range of tasks using more people. Crashing is rarely efficient but may be effective. It can only be effective if used for very short periods and if it is well-planned. Good planning, and the communication of those plans to the participants, are vital if crashing is to be effective.[64]

Needless to say, the work of time management must not only concentrate on the work on the critical path, but also pay close attention to work that is not now on the critical path but which could very easily move onto it.

Cost budgeting data should be used to check and confirm time progress. The use of 'S' curves, which predict amounts of expenditure, can give warning of something wrong in the time-management area.

Managing time deviations and reducing unfavourable ones is regarded as a central project management task.

Cost performance

By the time the subcontractors start, the project manager has budgets, or should have budgets, for the work. A control cycle of monitoring and applying pressure to get corrections begins.

In assessing the risk to a budget, the project manager should remember that one cannot apply control to the past, only to the future. Hence it is worthwhile concentrating on items that have a significant part of their budget left in terms of an 'estimate of cost to completion' rather than on those items for which most of the budget is spent and most of the work is done. In looking at the control issue, it is also worth separating out those aspects over which the project manager has some influence.

The quality of budget control is entirely dependent on the quality of estimating the value of the work done and of the work yet to be done to finish the job.

The project not only has to bear the cost of the work packages themselves, it also has to bear the cost of the interface work and of much of the common infrastructure. Often the main penalty for a delay in work will come to the project as an impact on infrastructure costs rather than work package costs (much of the risk of work package costs will have been transferred by contract to the subcontractor).

Motivation

The main responsibility for motivation of people working on the project sequence will actually lie with the subcontractors. Supplier project managers are able to have a significant influence on the motivation of their own teams but can only indirectly influence the subcontractors. Ensuring that the subcontractors themselves are not unduly stressed, ensuring proper production infrastructure is available, and keeping changes to a minimum are probably the most important strategies to maintaining motivation.

Contract administration

Each contract governing a project sequence will have a small number of key sections. Usually these will refer to procedures for payment, timing of work, change procedures, escalation provisions and dispute management.

Many of these procedures will have time deadlines and it is important for the main supplier to be set up so that these deadlines can be adhered to.

It is also important to note that certain records are vital and must be kept up-to-date. For instance, the date of receipt of instructions will be crucial if an application for an extension of time needs to be made.

Project sequence financing

In this area, the primary objective for the main supplier project manager is to turn the project sequence into a positive cash flow as soon as possible and keep it there, that is, a positive cash flow for the supplier organisation.

The project manager has to arrange for payment of the subcontractors and other charges. These are very often upfront charges and consequently represent outlays by the supplier. If the contract is well drawn up, the supplier may be entitled to higher rates of income at the early stages rather than in the later stages. This helps turn the cash flow positive.

Security of payment is a serious concern in project management. Clients can be very unstable organisations, eg, developers in the building industry; their tendency to be unstable is due to the 'lumpy' nature of the project business ('lumpy' in that their work comes in large lumps rather than being evenly spread out over a period of time) and by their high financial leverage.

The main supplier project manager should be working with a cash-flow projection and arranging for funds to be available in the accounts when required.

Payments to subcontractors

A system needs to be established whereby the work of the subcontractor is inspected and evaluated, and then payment made.

The danger for main suppliers is that they may accept work from a subcontractor and pay for it, only to have the client project manager reject the work and refuse payment. The nature of the contract will influence how this should be managed. The main supplier project manager should endeavour to obtain the client's acceptance of the subcontractor's work before making payment, particularly in the case of later payments. The client project manager, though, waits as long as possible before committing to acceptance of the work. This process could be the subject of negotiation between the project managers.

Non-performance

Most subcontractors want to do a good job—they want to come on the job, do it well, leave it and move on to the next. But now and again some subcontractors fail to perform satisfactorily; they may not come when required or the quality of their work may be unacceptable.

This poses quite difficult problems for the project manager. Replacing a subcontractor inevitably costs time—it certainly costs money due to the extra management time it consumes. If the work of the subcontractor is on the critical path, then replacing it costs the project time; if it is not on the critical path, while the project may not suffer an overall loss of time, people will.

The first step should be to look carefully at the critical path and see if the work of the subcontractor can be removed from it.

Before deciding to replace the subcontractor, the project manager needs to be quite clear as to the contractual position and know the costs involved in following a replacement strategy. With this clarification, the project manager then needs to examine whether it is absolutely necessary to replace the subcontractor. This process might involve the project manager in negotiations and attempts to persuade the subcontractor to lift their performance. The subcontractor's capacity and motivation should also be reviewed.

In entering this area of negotiation, the project manager might also like to consider his or her Best Alternative to a Negotiated Agreement, or BATNA.[65]

Control of subcontractors

This is not an issue of actual performance on the job; it has more to do with ensuring that the subcontractor not only turns up when expected but also stays on the job at the required level. Reasons for subcontractors not supplying may have to do with non-payment of a bill, other work of theirs which has been given a higher priority, or difficult supply conditions.

Temporary organisation

The implementation phase leads to a rapid increase in personnel working on the project. Many of these people are recruited through the subcontractors; some are in permanent employment with the subcontractors but many will be temporary. The main supplier will also experience an increase in employees.

Initially, there needs to be a way of checking that all employees are appropriately qualified. For this and other functions of a personnel system, a system needs to be established to deal with such matters as recruitment, terminations, qualifications, medical certificates and workers' compensation. Where the project uses people sourced from the international labour market, the personnel function becomes all the more complex and demanding.

As might be expected, personnel requirements vary from project to project. There should be an explicit plan to deal with the build-up and reduction of the project organisation.

The form of the temporary organisation on many projects is a central organisation under the direct control of the main supplier project manager, to which are connected various subcontract organisations. For internal clients, such as one's own organisation, the structure might be a loosely affiliated task force.

Materials handling system

The movement of materials is a big component of many projects; on some, it is the key determining issue. Materials handling is a major consideration in the organisation of pop concerts, building projects, office relocation projects, projects introducing new products, and so on. Materials handling has become a specialist area with a very wide range of equipment available.

Setting up and running storage systems and waste management systems are part of the materials handling issue.

The installation and removal of a materials handling system is usually not shown as a program activity for the client, but it must be shown as an activity for the main supplier project manager.

Organising materials handling may also involve negotiating with the police and local authorities.

Industrial safety

Significant legal obligations are developing in this area. Work has to be carried out in a safe manner and the obligation to ensure this happens rests largely with the employer. The legal obligations are quite onerous and in many case reverse the onus of proof, from 'innocent until proved guilty' to 'guilty until proved innocent'.

Managing industrial safety has become a specialist activity. The ideas behind the management of industrial safety are very close to those supporting TQM. Concepts of systems of work, capabilities of systems and so on, are very relevant.

Project managers must arrange a safe system of work for work directly under their control, and also ensure that each subcontractor has a safe system. A safe system of work is usually planned and described in a work method statement, which describes how the work is to be done, what equipment is to be used, what special precautions need to be taken, details of how to handle chemicals and materials and the qualifications required of personnel; it also refers to the relevant codes and standards. Failure to ensure these safe systems will leave the project manager seriously exposed to legal (which will probably be criminal rather than civil) action.

Industrial relations and the project sequence

Most project managers see this as a difficult area, and it is. The terms and conditions of employment are an important ingredient and quite often the main supplier organisation is involved in setting out some key parameters governing them. However, because it's a difficult area is no reason to ignore it; it should be approached in a rational manner.

The terms and conditions of employment have to be decided. Most of these are agreed between the subcontractors and their employees before joining the project sequence and therefore most of the decisions are outside the influence of the project manager. But sometimes there are project-wide agreements which have to be negotiated by the main supplier project manager.

There are two elements to industrial relations. On the one hand, there are the rules, conventions and laws within which industrial relations activity operates, while on the other hand, there are politics.

The rules, conventions and laws governing industrial relations influence how people organise themselves and behave. They apply to employer organisations, unions, awards, enterprise bargaining, strikes, various judicial systems, and so on. To deal with the rules part, the project manager needs to appoint someone or have access to someone who knows the rules, conventions and laws. These rules are quite powerful definers of the limits of industrial practice and behaviour, and knowing them in some detail can be very useful when involved in negotiations or in some difficult processes such as terminating employment.

The politics are more difficult. The politics associated with overall social standards are one part of the political agenda and are essentially outside the influence of project managers. What project managers should know, however, and make their business to know, are the political leanings and persuasions of the participants on the project. These are often the bases of the values participants bring to negotiations.

The project manager needs to see industrial relations practice as part of the overall management strategy. The conduct of industrial relations influences motivation and morale. It does so in ways that are difficult to measure but which many believe have an impact. Its impact will be on productivity and quality—and also on how much support an employer gets when something really difficult occurs, such as an accident.

Applications for approval of variations

This involves establishing the appropriate price for a variation as well as the appropriate time allowance. Justification for the proposed change is usually needed also. Where the change has been initiated by the client, the main supplier project manager does not need to provide a justification, just cost and time, with whatever other qualifications are appropriate.

Just as the client project manager needs to control the flow of variations, it is necessary for the main supplier project manager to do likewise. The main supplier has to run its own schedule of changes and record their status. The main supplier's list of variations does not include proposed client variations which did not gain approval.

The main supplier has to know that whoever on the client side is issuing instructions to carry out variations also has the authority to so direct the main supplier and bind the client. The main supplier also needs to know that the client can pay for it.

The main supplier also has to manage variations that arise between the subcontractors and the main supplier that may not involve the client.

The accounting system

Each project will require its own accounting system to track its transactions. This tracking will cover such items as supply of funds to the project, the payment system, procedures for the authorisation of payment, purchase orders and account numbers. On any reasonably large project, this work should be done by specialists in accounting. Assessing one's financial status with projects can be quite difficult because of problems in assessing the value of work in progress.

The documentation system

This is essential. The documentation system and processes have to be set up and, equally importantly, be maintained. Among the documents of interest in the implementation phase might be drawings, specifications, orders, invoices, part numbering, instructions, variations, distribution requirements, inventories, work in progress reports, acceptance or rejection of work, photographs, completion notices, inspection reports, repair reports, time sheets, cost reports, time reports, weather reports, deviation register and corrections, start dates, personnel records, travel expenses, pay records, debit notices and invoices. The range and number of documents on projects can be quite extraordinary.

The real difficulty with setting up and managing a documentation system is to organise the collection and storage of data in such a way that the data is easily accessible.

Managing external functions: the client

The client external functions are managed by the client project manager. The opportunities for and threats to the project lie mainly in the external environment, hence it deserves special attention from the client project manager. Many of the areas to be managed in relation to the external environment interact with one another, so there will be some duplication in the descriptions below.

Maintain clarity

It is imperative at this stage of the project life cycle that the client organisation projects a clear message as to where the project sequence is going. In small organisations, this may be easy, but when large organisations are acquiring projects, keeping a clear message is quite difficult.

As the project is proceeding, there are many who still want to change its content. So the project manager still needs to keep an eye on stakeholders and various others to ensure that unnecessary massive change does not occur or, if they are to occur, then they do so in an orderly and controlled way.

Maintain enthusiasm

Maintaining enthusiasm for the project means promoting good news about the project and its project sequence. It also means dealing with difficult and unfavourable publicity.

The project manager will need to identify key players outside the project and make sure they are well-informed and see and promote the benefits of the project. This requires considerable effort.

Clear ownership

Maintaining ownership involves much symbolic action, action affirming ownership and action that continues to claim ownership.

Confusion of ownership is a threat to a project. The project manager needs to keep a clear line of ownership; the loss of this clear line leads to confusion, faulty decision-making and massive delays.

An example of where ownership of a project changed hands occurred on the Sydney Opera House project. While it was under construction, debate about its function continued—there was a clash between those who supported opera and those who supported orchestral concerts. For technical reasons, mainly concerned with reverberation times, opera and concerts are rarely presented in the same hall. Thus, in a building with two halls, one was designed for opera and the other for concerts. Initially, the bigger hall was designed for opera with the appropriate stage machinery. The balance of power shifted and those supporting concerts got the upper hand, leading to a decision to put the opera in the smaller hall.

An acceptable public profile for the project

Many projects operate in publicly sensitive areas. Management of these areas usually involves trying to get the project story across. It needs to counter misinformation (in some conflicts, it may be hard to distinguish between information and propaganda) and replace the 'opposing' story with an appropriate version of the project story. This often becomes the province of public relations practitioners.

Somebody on the project management team, or the project manager, must deal with this issue. It is crucially important that energy be devoted to dealing with the external environment, to dealing with the external stakeholders and groups who have the ability to affect the project. The internal environment also requires attention but this is usually so obvious that attention is directed there anyway, often at the expense of the external environment.

The continued supply of resources

Now that the project is up and running, support must be maintained, especially from those who supply the project with resources. These external resource suppliers include financiers, other departments and other organisations, which might be government or private. There are many examples of projects losing political support in key organisations and the project having to be abandoned as a result. Many people need continual persuasion and reassurance; for example, so that finance continues to be advanced, time may have to be spent reassuring bankers of the project's progress.

Keeping key stakeholders informed consumes a great deal of time, especially the time of those at a high level—in fact, only the high-level people are likely to have access to those who influence supply. Project managers need to ensure that they have access to the key people, to boards of directors and politicians.

The project manager might consider setting up formal links and structures to be used to protect the project. One way of being proactive is to set up panels whose purpose, ostensibly, is to oversee the project progress but who, in fact, provide considerable protection for the project.

The need to exercise power and influence increases with the length and complexity of the implementation phase. Short implementation phases are probably over before supporters can change their positions. However, a long implementation phase implies a need to set up contacts, structures and coalitions to support the project.

Some project managers regard maintaining support as so important that they leave the rest of the management of the implementation phase to others so that they can concentrate on it.

Continued support for the client project manager personally
Project managers must look to their own positions. Comments made above in relation to maintaining support for the project also apply to project managers protecting themselves.

Managing external functions: the main supplier
The main supplier external functions are managed by the main supplier project manager, who must ensure that required resources are available in the market-place and that those who are to receive the project are willing and ready to do so (thereby helping to assure payment).

The market can supply
The main supplier has the responsibility to source all (or most) of the requirements for the project sequence.

The receivers of the project
A successful project requires that the users take on the project when it is completed. The client needs assurances that the project is acceptable to the users, who may need training and other forms of introduction to the project.

Managing the implementation phase

The implementation phase has been described along with some of the management issues. Besides making sure that all these activities are taking place, the client project manager has to keep an overview of the project sequence. The project manager has to identify all the parties, bring them together as appropriate, allocate responsibilities, develop overall co-ordination, conduct meetings, remove road-blocks, resolve conflict, set priorities, prepare people for handover—all the while keeping an eye out for a sudden need to dramatically change direction. (A dramatic change should only occur when the external environment changes, but sometimes a major mistake is found in the work justifying the project, or else a sudden failure occurs.)

Much of this work involves knowing where the overall project is at, and then using various social processes of persuasion (and sometimes raw power) to keep things on track. Most of this is done by talking to people and through the effective conduct of meetings.

Talking and listening

Oral communication is a central project management activity. While some of the work of project management involves writing such things as progress reports and recommendations, most of a project manager's time is spent talking. The wider the group of project participants project managers talk to (and listens to), the better informed the managers will be. Paperwork is usually produced only to confirm what has already been discussed.

Holding meetings

A series of parallel meetings are held during the implementation phase. The client continues to hold meetings with key stakeholders, and the client project manager holds meetings with client representatives and project supplier representatives. Each of the project suppliers will have their own meetings.

Project Control Group

At the client meetings with key stakeholders (usually called the Project Control Group, or PCG), reports on project progress are reviewed and decisions made concerning corrective action and project changes. Those attending these meetings usually expect to have to respond to recommendations. The person in the chair is usually a senior executive of the client organisation. These are important meetings to the project manager in that the people at it are more likely to be close to any changes in the environment that are likely to affect the project. Meetings need to be managed by the project manager in such a way that confidence in the project manager personally, as well as confidence in the project, is maintained. Problems brought to these meetings, or issues requiring resolution, need to be presented with recommended solutions or ways of dealing with them. At a minimum, there should be a recommendation as to who should be delegated to investigate the matter.

PCG meetings can be used by the project manager to exert pressure on some key stakeholders who are not performing or who are causing problems for the project.

Site meetings

At the next level are meetings usually chaired by the client project manager—here, they are called site meetings (although they need not be held on a building site). Those usually attending are the main supplier and client representatives. Where there is more than one contractor, each contractor usually sends a representative. Some meetings at this level involve the client representatives, the contractors and also their subcontractors; this is more unusual but is quite useful to the client project manager in keeping a real hold on the details of the project.

The meetings usually involve reporting on the progress of work, setting out the planned work for the coming period, and highlighting problems that exist now or are likely to come up. At site meetings, the client project manager seeks commitments from the various participants to achieve various stages of their work and confirmation of the detail work ahead.

In the implementation phase, the purpose of co-ordination is to carry out the work as planned. Assuming that the work has been well-planned and that all the participants are carrying out their work as expected, site meetings probably perform a monitoring role. In practice, even well-planned projects have problems with things not going as expected, or with some new item being introduced.

Where there is considerable uncertainty in the planning or where things are very noticeably not going to plan, site meetings play a key role and need to be managed very tightly. The project manager should have already decided what the outcome of the meeting should be: making that clear helps the meeting to progress and helps maintain people's commitment to the project.

Project implementation meetings are often much more diverse than those held during the design stages. During those stages, meetings consist of expert consultants who all have the same or similar status. Site meetings have people from many walks of life and from many different socio-economic groups. The project manager needs to handle the meeting so that people will feel it easy to participate.

Again, things are not decided by a vote at site meetings, they are decided by an agreement that what has been put forward is in accordance with what is expected as a result of the contractual relationships. There is a common motivation to get on with the job, finish up and leave with a job well done and some profit. So these meetings have basic motivating processes. They are essentially a co-ordinating device used to expand upon and put into operation the contract requirements. Particular problems or arguments about what is agreed are usually handled outside site meetings.

Contractors' meetings

These are in many ways similar to site meetings but they deal with much more of the nitty-gritty, covering detailed supply issues, resourcing, facilities and payments. The results of these meetings are fed into the site meetings. Contractors' meetings take place within the framework of quite clear legal agreements, so directions can be more easily given. None the less, these meetings have an element of agreement and have an impact on morale and need to be managed appropriately.

The next phase

The last part of the implementation phase is handing over the project to the users or investors. This last part has come to be recognised as a phase in itself, worthy of its own separate management. In fact, this phase needs to be managed as a subproject working to achieve the project sequence end-point. Attention is now turned to the handover phase.

Chapter Eight

Handover Phase

The client project manager and the main supplier project manager manage the handover phase together.

The main purpose of the handover phase is to transfer the project in a functioning form from the main supplier to the client. A single project that includes physical equipment, organisational changes and information system changes, also includes that equipment and those changes actually working. Making a thing and then just leaving it can be a project but it is unlikely to be a success (unless it is a statue). The project is the final state of being after handover. Hence the need for a full and proper handover phase, the last step in the project sequence where everything is put together, in working order, and placed in the hands of the client.[66]

Two terms are usually encountered here: 'commissioning' and 'handover'. Commissioning normally applies to technical projects and getting them up and running, while handover is the transfer of legal title to the project. Some expect the project to be commissioned before handover while others expect handover and then commissioning. This conflict in expectations is quite reasonable given the nature of the phase. The term 'handover' as used in this book covers the whole sequence from the finishing-off of implementation to the actual functioning of the project under the client's control.

This phase does not have a neatly defined beginning and end. It starts somewhere during implementation and ends at a point where the client is deemed to have full control and responsibility for the project. While the beginning is never precise, some legal fictions are introduced into project contracts to create a definite end-point. On many construction projects, a date of 'practical completion' is certified by the client's representative; completion is then deemed to occur 12 months after that date (it is not quite as simple as this, but it is an example of an attempt to define an end-point). On such projects, the main supplier project manager has a clear target in this phase to achieve the certificate of practical completion.

The client has conflicting interests in this phase. On the one hand, the desire to benefit from using the project encourages the client to accept it as soon as possible, but, on the other hand, the client is inclined to delay until assured that all is satisfactory and all the contract conditions have been complied with.

The main supplier's interest is probably less complicated, unless reputation is an important ingredient in the outcome. The supplier is driven to complete handover by the desire to be paid as soon as possible. The supplier also wants to achieve handover with the minimum of ongoing liabilities, and the desire to minimise these liabilities may act as a deterrent in expediting the handover process. Many suppliers, however, ignore this and deal with ongoing problems later from general revenue rather than from project revenue.

All the work of handover needs careful planning. Some quite difficult methodological issues have to be sorted out in testing various technical parts of the project and in ensuring ownership has been effectively transferred.

The handover phase is a finishing-off task, it is closing off the details and, in many cases, dotting the 'i's' and crossing the 't's'. Thus it needs a careful, methodical mind, not one swept away with possibilities and opportunities. It needs someone who is committed to tidying up everything, a rather different mind-set from that required in the earlier phases when all was possibilities and there was the drive to get things moving. There is probably room for specialist commissioning project managers.

The work of handover

The work of handover consists of ensuring that the contractual conditions have been satisfied, that what has been delivered conforms with specifications, that the project is integrated into the ongoing business, that transfer of legal and psychological ownership occurs, and that all the bills are paid.

Handover relies on the work of the planning and design phase to give direction and guidance. Designers and planners must be aware of this need to define their requirements in such a way that acceptance can be based on measurable criteria; failing measurable criteria, then based on identifiable processes or sequences.

Ensuring contractual completion

The contract provides the overall framework for handover. Where it is clear and where the project sequence has worked well, there is only the problem of ensuring that the technical requirements are satisfied; this can, of course, be quite difficult.

Where the contract is unclear or where the project sequence has encountered difficulties, the project managers have to find a way through. Failure to take the initiative here leads to a real possibility of useless, frustrating and quite damaging delays occurring because decision-making becomes paralysed. If the contract conditions cannot be met, either because the conditions are unclear or because of technical failure on the part of the supplier, there are a number of approaches the project managers could consider. They could try and resolve the matter:

- by one side accepting the contractual consequences;
- by negotiation;
- by use of arbitration;
- by use of commercial conciliation;
- by going to the courts; or
- by finding a special contractual condition that allows resolution.

Accepting the contractual consequences may imply paying penalties as laid down in the contract. Negotiation should be conducted in the framework of an ongoing overall handover negotiation to allow appropriate give-and-take opportunities to arise. Some arbitration processes can have the force of law; conciliation often does not close off this option.

In such situations, the project managers must form judgements as to the likely outcome of each of the above approaches. This clearly means that some preparation work needs to be done and a decision made—there will be no perfect answer. It is worth remembering that a commercial decision has to be made, not a perfect technical decision.

Clearly, the decision-making authority must be known. The contract may say who has the authority to sign agreements, and thus make a commitment, but these people may not have the real authority to make the decision. Thus each side has to ensure they are getting through to the people with the full authority to make the decisions.

The contract document itself should have in it the mechanism by which completion is agreed or can be identified.

Structure of contracts

While the client side may be involved in a small number of project contracts, the main supplier is involved in a wide range of subcontracts. Each of these has to be closed and finalised, so it is not just the closing of one contract that is involved here but the closing of a whole structure of them. It is necessary to work out a path through these closures; some contract closures will be necessary prerequisites to others.

To allow some of these contracts to close, certificates of various kinds are required as evidence that, as far as the other side is concerned, the work has been completed as required by the contract. So project documentation includes a range of certificates.

Difficulties for the client can arise when there is failure to perform in one or more of the smaller, but critical, contracts. Where a client has a contract with a main supplier who in turn has a contract with a subcontractor, the main supplier may be seriously hampered if the subcontractor is a non-performer. The legal rights and structure of obligations may be quite clear, but waiting to work through the full legal process may not be in the client's interests.

Retentions

On many contracts, there is a provision for the client to withhold a small percentage, often around five per cent, of the contract sum and to pay it at a later date. This amount is called a retention, and is held against the possibility of defects being found in the work after the work has been completed. It is the main supplier project manager's job to try to reduce the amount of retention; the client project manager tries to do the opposite.

Maintenance contracts

For many projects, the delivery of the project may only be the start of the most important part of the project—maintaining the performance of the item sold as the project. Maintenance contracts are often treated formally as separate contracts, one for the supply of the project and the other for its maintenance. Many computer systems and products, both hardware and software, are supplied with the clear need for ongoing maintenance.

Operational documentation

Some project documentation is notoriously difficult to obtain. There is often great reluctance to produce operating manuals and as-built drawings; there is an idea that the physical equipment side of the project is the end in itself and that operation is a minor matter. Work on the documentation needs to start quite early, certainly before the end of implementation. Getting the documentation to a satisfactory standard usually involves ongoing negotiation.

Certification and warranties

Certificates may simply state that all is satisfactory from the buyer's point of view. However, there are other quite significant certificates, particularly those that affect the legal operation of the project. Some of the warranties may have significant technical content, which can be particularly important when incorrect operation could invalidate a warranty.

Conforming to specification

Conformance to specification is established by monitoring the mode of production, by testing what is finally produced, or by both monitoring and testing.

Monitoring and testing are often both carried out on large contracts. Major airlines monitor the production of the planes they have ordered and then test them for performance when completed.

On a misty morning in 1967, six panzers rolled onto the Pfaffensteiner Bridge in Regensburg, Bavaria, Germany, and took up position. The survey team got to work and measured the deflections of the bridge under the load of the tanks. Fortunately the deflection was within the 20 mm limit required by the specification and the bridge was deemed to conform. This test was carried out despite the fact that the whole sequence of the production of the bridge had been carefully monitored right from the start.

Where conformance to specification is based on the method of production alone, the acceptability of the work has to be decided as implementation proceeds.

It is actually very difficult to prove that no fault exists in the work completed—it is probably theoretically impossible. Anyone who has tried to debug a computer program understands this difficulty. What one does, in fact, is reduce the possibility of failure, not eliminate it.

The strategy advocated in dealing with this issue of accepting technical equipment is to break the equipment into parts, examine the operation of each part, and then examine the interactions of those parts. For each of the parts, measurable performance criteria should have been developed, preferably in the planning and design of the units or parts before testing.

Conducting acceptance tests

Most projects require acceptance tests. On projects such as organisational change projects and political campaigns it might be difficult to identify acceptance tests. These projects may be able to be tested for feasibility of certain strategies but acceptance tests as such are impossible. There are other projects, such as the introduction of new products, where the test comes in the actual use of the project; acceptance tests may have been carried on the production equipment but the important test will be market acceptance. This test can only be carried out by doing it.

Any testing program beyond the simplest requires a plan. Some tests are so complex that they might need to be treated as subprojects with their own project managers and resources. The conduct of acceptance tests is a crucial step and, in selling terms, is an important part of the 'post-selling' of the project.

Nomination of tests

Since the nature of the acceptance tests depends on the specifications, the development of the specifications should include methods by which conformance can be recognised. Vague statements calling for perfection only delay the work of handover. The acceptance tests should either be clearly specified in the contractual specification or agreed between the supplier and the client. Relying on agreement to be developed between the supplier and client is problematical if no clear guidelines can be identified.

However, some level of new agreement will have to be developed between the client and the supplier because no specification can cover all possibilities; there might even be conflicts within the specification, and there may even be work now regarded as being no longer required or superfluous.

The acceptance tests may in fact be determined by a third party, such as a government agency, because of safety or environmental concerns.

Specialist involvement

Acceptance tests involve specialist staff from both sides and may also involve a range of third party suppliers. The task of getting the specialists and the required equipment together is best served by an agreed program for the conduct of the tests. When such a program is drawn up, it should take into consideration not only the technical requirements but also the availability of personnel.

It is essential that those present have the authority to sign-off as required. These authorities may range from the authority to sign that a result has been observed, to signing actual acceptance of the test result.

Rehearsals

For some tests this is clearly impossible, such as those requiring large amounts of resources. Test rehearsals are a useful way for the supplier to increase the chances that the tests will proceed smoothly. While tests can help to build up the client's confidence, the client needs to be wary and vigilant. Computer demonstrations are notorious for their ability to do amazing feats while, hidden from the view of onlookers, is the huge volume of work needed to set up the demonstration.

Client responsibilities

The acceptance procedures for technical equipment may involve the client in providing a range of goods and services. There have been problems on projects where the client has not been able to provide for the tests. An example might be lack of feedstock for a trial run of a chemical plant, or of a particular grade of coal for a coal washery acceptance test.

There can also be some unintended consequences of these trial runs. For the trial run of the plastics section of a car factory, it was decided to produce garbage bins; this resulted in a huge oversupply of bins which almost wiped out the local garbage bin manufacturers.

Extent and duration

Some trial runs last weeks—it is not unusual for parts of a production plant to be run continually for periods of hundreds of hours as part of an acceptance test.

Covering all eventualities may be impossible. It may take far too long or be quite difficult to do. Here there is the need to set priorities; these are, of course, specialist decisions, not decisions of the project manager. The project manager may point to the need to set priorities but actually doing it depends on the specialists.

Recording the tests

The tests are conducted as required by the technology and/or the contract. But it is not just the conduct of the tests that is important; records must be kept so that certificates can be prepared on the basis of their contents. Accurate, complete records are vital so that, in the event of a failure or a disputed test result, the data is available for examination and review. Some of the records may have to be tendered in court. The records should describe the tests, their conduct, any deviations or variations, and so on.

Failed tests

Acceptance tests may, of course, expose lack of conformance. This may or may not be essential to the project; more critically, it may or may not be capable of being brought into conformance. Resolving these problems requires negotiation or another procedure, which may simply be the acceptance of some penalty set in the contract.

Failed tests create dilemmas for both the client and the supplier. Both the client project manager and the main supplier project manager are obviously embarrassed. If there is to be a failure, it is best that it be in private. Public humiliation leads to loss of confidence.

For important parts of the project, contingency plans should be prepared so that there is an agreed procedure on how to handle the problem of failed tests.

Successful outcomes

Successful acceptance tests end with the issuing of an acceptance certificate by the client to the main supplier.

Integrating the project into existing operations

The success of the project is partly, and significantly, dependent on how the users take to it and use it. The project has to be transferred in terms not only of technical operation but also in the more subtle ways associated with ownership. Failure at either level detracts from the success of the project.

Introducing a new piece of very expensive equipment into a hospital requires not only a range of technical infrastructure adjustments (power, lighting, air-conditioning, structural support and health protection devices), but also a range of personnel training and procedures to deal with how the new equipment alters the operation of the hospital. Satisfactory technical performance of the new piece of equipment is only part of the concern of handover.

There are some problems here with definitions in relation to the word 'project'. To deal with them, the acronym POIPH (pronounced 'poif') which stands for Product of Implementation Phase is introduced. The problems arise because it was stated at the beginning of this chapter that 'The project is the final state of being after handover.', a statement entirely consistent with the original definition. In other words, the project includes the completion of the integration (of whatever, a plant, an organisational change, a new product) into existing operations. Therefore to say that the project is to be integrated into existing operations is logically inconsistent, as such a state is already included in the definition of the word 'project'. To overcome this inconsistency, POIPH is introduced, so that is now the POIPH that must be integrated into existing operations to complete the project.

This last piece of integration (of the POIPH into the existing operations) is often not seen as the concern of the project manager. Thus, many projects, by virtue of their definition (ie, by their failure to distinguish the POIPH from the project), are almost set up for failure. It is an interesting example of how inadequate language or definition can blur important issues.

The form of the integration of the POIPH into existing operations should be decided during planning and design, but often it is not. In fact, the integration is usually subject to considerable negotiation towards the end of implementation and during handover. What project managers have to do is push and negotiate events, and get people to agree to and conform to an appropriate integration.

The exact shape the project finally takes has to be negotiated with the users. Introducing the users to the POIPH and its operation may involve training, visiting other establishments with the same project, presentations, newsletters, researching and responding to user concerns, and providing ongoing support as handover proceeds. If the use of the

POIPH represents a big change to an organisation, it may be useful to refer to the literature on organisational change.

There is of course a political dimension to introducing a new POIPH into the operation of an organisation. The political dimension also implies the use of power, in that the POIPH my be imposed on the users. Part of the skill of the political operation here is to know how to get sufficient support to persuade the opposition not to raise objections.

Other balancing activities need to be undertaken. The users must be encouraged to look forward to their role of ownership of the project, to increase their commitment to it and the project's chances of success. Many of the users and operational stakeholders must therefore be identified and accommodated; all of these various people will have different interests which need to addressed. This is an area where industrial relations often comes into the project manager's orbit.

Clearly one does not have to wait until the end of the implementation phase before bringing the users on board. What is important is that, by the end of handover, the client or owner has successfully taken over the operation of the facility.

Meeting legislative requirements

Compliance with legislative requirements is a fundamental condition to satisfy on those projects where such requirements apply. The project managers needs to identify if such conditions apply and integrate the requirements of compliance into the project sequence.

Tidying up

There is the tidying up or closing-off of all the procedures and processes associated with the project sequence. This involves making sure that all outstanding approvals are either given or refused and, if refused, taken to a conclusion which may involve preparing for and running a legal action.

There is also the tidying up involved with making repairs and fixing what is wrong. This is usually formally dealt with by agreed lists of repairs or corrections to be carried out. These lists go by various names, for example, deficiency list, punch list and defect list—the names vary from industry to industry. These lists are often made on a formal inspection of the POIPH. The main supplier then has to go through and make the changes or repairs which are subsequently inspected and, if satisfactory, are deleted from the list.

Making final payments

All the bills have to be paid. Accounts have to be closed, audited as needed and the final accounts prepared.

Final payments are important matters and, from the client's point of view, represent a significant reduction of power over the supplier. Equally, for the supplier project manager, the final payment represents the final step in disengaging from the project and the client.

Before final payment is made, the client project manager needs to identify what has to be satisfied beforehand. The payment is usually paid in response to an invoice from the supplier. The basis of this invoice is examined by the client project manager and advice prepared. If the advice is to pay the account, the client usually complies. If payment is to be withheld, or partly withheld, the client project manager needs to ensure that such a step is within the contractual power of the client.

The work of commissioning

The brief description of commissioning above outlines what is required to hand over the project. But in order to get this work done, other things have to be accomplished. Some of these have been referred to, such as making arrangements for tests, agreeing on schedules, supervision and getting advice.

There is also work which is not directly implied but is always there, such as conforming to safety regulations, paying people for their work (for instance, paying the accountants who pay the main supplier), disposing of equipment and material, preparing the necessary records, archiving material, preparing the accounts to be integrated into the company accounts, and cleaning up.

What should also be included is a feedback process so that experience gained can be used on future projects. But, for practical reasons, this is quite difficult—people move away from a project and are hard to contact. There is also a theoretical reason why it is difficult, in that it is not yet clear what the feedback should entail. Collecting data on projects is still a difficult methodological issue.

There is also work involving people. A morale problem has to be dealt with at handover. Handover is for some a sad and uncertain time; some people are quite fearful and are intent on prolonging things. Others, however, while feeling a bit sad, have lost interest and are now focused on the next project. This human side requires skilled human relations management to bring the project sequence to a smooth conclusion.

Managing the work of commissioning

Managing this phase means planning the work mentioned above, initiating it, adjusting it and keeping it going until the project is complete. The management includes setting up sequences or processes to ensure contractual completion, to handle the interfaces between the various elements contributing to handover, to handle disputes, to run acceptance tests, to integrate operations into existing operations, to tidy up a whole range of matters and to arrange final payments. It requires:

- identification of the work (use project start-up meetings and a work breakdown structure);
- identification of relationships (use project start-up meetings and a network);
- estimation of work-loads;
- allocation of responsibilities (use a responsibility matrix and an organisational chart); and
- protection of project morale (use open systems of reassignment and keep people up-to-date).

This management is actually conducted by the client project manager and the main supplier project manager without each fully recognising their mutual dependence. The process requires either explicit or implicit agreement between the client and the supplier, and sometimes with the users.

On many projects, this is the first time the users are introduced to the project—often for very good reason. Often, the users only come together or are selected towards the end of the project sequence. Part of the handover may include the appointment of the users, thus making it difficult for them to participate in it.

Handover is two or more subprojects

For each side, handover represents a subproject. If the client and the main supplier are only involved, then there are two subprojects. However, more are usually involved, the client, the main supplier, the subcontractors to the main supplier, and possibly the users. Each side has a subproject to manage, although they are intimately linked together. The management of each subproject can be divided into phases, the more useful ones being the planning phase and the implementation phase of each of the handover subprojects.

The more useful aspect of applying the idea of a project or subproject to the handover phase is that the work can be reframed into new and separate projects, thereby drawing on much of the morale power of involvement with projects.

On most projects, a lot of the work of handover actually runs in parallel with the implementation phase. Thus it is useful for the main participants to give people the roles of subproject managers with the prime responsibility of completing handover. This is a useful device to allow a changeover of project managers before the end of handover. In many cases, the principal project managers are transferred to new projects while still having responsibility for finishing off the last project. Handover becomes a low priority in this environment.

The subprojects provide a focus and need their own resources and timeframe.

Disputes resolution

This is the time to finish the arguments. To deal with arguments requires a disputes resolution process. The project managers have to appoint people with adequate authority to deal with the contentious issues, and each has to prepare for handling them. There should be an agreed procedure and hierarchy of dispute resolution processes between the various project mangers, or subproject managers. Having made their preparations, the parties start the processes, the first of which might be negotiation. Failing that, the argument may proceed to commercial conflict resolution and, failing that, the courts.

A lot of new disputes can come into the project with the appointment of the users and in trying to hand the project on to them. It is important for the project managers to curtail new demands.

Archiving and recording

The final step in handover will probably be the archiving of the project sequence records. Not much of the feeling and passion of the project will survive in the records, thus much is actually lost in terms of learning from experience on other projects.

Chapter Nine

Project Sequence Capability

Three concerns, although related, can be isolated and presented as questions. First, will the project as proposed satisfy the objectives? Second, will the project sequence deliver the proposed project? Third, will each phase deliver the output that is expected of it?

The answers to these questions are quite difficult to obtain. To answer them, a person, on the one hand, uses what is called 'capability', which describes the *expected* output, and, on the other hand, uses the definition of the *required* output. In so far as the capability overlaps the required output, the work is a success. In so far as the capability of the project overlaps the requirements of the objectives, the project is a success. In so far as the capability of the project sequence overlaps the proposed project, the project sequence is a success and, in so far as the capability of the phase overlaps the expected output, the phase is a success. Although it may not be logically necessary, we are concerned to organise a successful phase which contributes to a successful project sequence which in turn contributes to a successful project.

While the required output may be easier to define, and this has been dealt with earlier within the context of the threshold of complaint, the capability will almost always be measured only imprecisely. But even this imprecision is useful, and leads to a vast improvement in comparison with operating without any sense of the capability. Thus we are interested in developing a feel or sense of the capability and assessing the degree of overlap. Then the managerial action is to increase the degree of reliability and of overlap. Most of the work described so far, the work through the project life cycle, has been directed towards obtaining an overlap between the capability and the objectives.

The idea of capability, as presented here, has its origin in the quality literature. It owes its origins to ideas developed in the management of processes, referred to towards the end of this chapter, where one will find a definition that contains considerable mathematical rigour. This rigour is too ambitious for one-off problems to be dealt with in project management so we will introduce looser definitions to provide a framework for our conceptual thinking.

It should also be noted that in the management of repeating processes, one stabilises the process and then tries to move the process capability closer to the desired output. We will not have much opportunity to stabilise the sequence and often will be engaged in both stabilising the sequence and moving it to the desired outcome at the same time. Thus a whole level of opportunity for finesse is not available to the project manager. In doing this, very crude methods are relied on, often totally dependent on the judgement of the project manager (this is not to be taken to imply that the judgement of the project manager lacks finesse!).

In trying to improve the capability of the sequence, we will be also trying to target it to the desired outcomes. In fact, much of the capability of the project sequence is built into the cybernetic model, the model underlying the control framework outlined earlier. This amalgamation of the ideas of capability and of desired output should not prevent us keeping the ideas separate and managing them as separate items.

In very simple terms, we are concerned to identify the objectives and then set up and manage a sequence capable of delivering those objectives. This is the essence of the description of the work of the project life cycle.

Capability

Three capabilities are defined below, each relating to one of the above three questions. They are quite separate issues, often ending under the control of separate people. The project manager needs to ensure that they are compatible.

Project capability

Project capability is a measure of the reliability of what the project can and cannot do. If the project capability coincides with or delivers the objectives, then the project will be judged a success. The project is produced by the project sequence. Project capability is not concerned with how the project is brought into existence but with what the project can do or achieve once it has been brought into existence. One might produce the wrong thing very reliably and accurately—this we wish to avoid.

Project sequence capability

Project sequence capability is a measure of what can be reliably produced or achieved by the project sequence. As the project sequence proceeds, good project management will be concerned to align the project sequence capability with the desired project and also to align the desired project with the objectives. Project capability and project sequence capability always need to be under review in project management.

Phase capability

Phase capability is a measure of what can be reliably produced or achieved by the work of the phase. It will be dependent upon the capacity of the production infrastructure available in that phase. Project sequence capability is made up or achieved by phase capabilities.

The parallel feasibility phase– looking after capability and objectives

The issue of project capability cannot be put aside or assumed to be without problems. The environment may have changed, the original project objectives may have been wrong, or the manifestation chosen for the project may be wrong. Thus while the project manager is managing the project sequence remaining after the feasibility phase, he or she must all the time monitor and control the overlap of the desired outcomes with the project capability and project sequence capability. This monitoring is equivalent to a parallel feasibility phase, parallel to the project sequence. This parallel feasibility phase is part of the project management.

The parallel feasibility phase involves environmental monitoring, examination of objectives, and checking on the viability; it also involves further elaboration and checking of the financial plan, the marketing plan, the technology plan, the quality plan, the production plan, the contractual strategy, the time and cost plan, etc. It involves being concerned with objectives, resourcing of the project sequence and the establishment of the infrastructure. It is the function of the parallel feasibility phase to ensure that the whole project sequence hangs together as a coherent and feasible whole and that it will produce a project which will satisfy the objectives. The function of the parallel feasibility phase is to develop, enhance and ensure the reliability of the project sequence and to increase the overlap of project capability and project sequence capability with the desired outcomes.

The project manager, in managing the project sequence from the end of the feasibility phase until the end of handover, must maintain a planning, acting and control function aimed at clarifying and confirming the objectives, and also aimed at enhancing the project capability and the project sequence capability. This is the work of the parallel feasibility phase.

For ease of presentation, the term 'project sequence capability' will incorporate 'project capability' and 'phase capability'.

Capability and its measurement

On projects, there are three possibilities. One of them corresponds to the typical process problem and quite well-known mathematical procedures can be applied; one of them corresponds to the one-off; and the other is a mixture.

Firstly, there will be activities which have no repetition and no history to look back on and use as a basis for assessment. In these cases, a theoretical model must be built up, almost from first principles, of what might happen. The one-off nature of projects means that this will have to be done at least once and a judgement formed using quite crude techniques. Despite their crudeness, they will contribute to surprisingly effective levels of control. For instance, the proposition that if there is no slack in the planned allocation of resources then it is almost certain that the activities will take longer and/or cost more than expected, can be used to check one's sense of confidence in the sequence. However, in a situation of no repetition, capability is probably most difficult to measure.

Secondly, there are activities that are not repeated on this project but have been done on other projects. Typically, this arises when a specialist subcontractor is employed who has done the work before on other projects. Here, a record of activity and output exists but this record is often not available (the subcontractor has it!). Where organisations work together over a long period of time, this can be a significant source of data for both the main supplier and the subcontractor. With appropriate agreements, some quite good measurements of capability should be possible here.

Thirdly, there is an activity that will be repeated on the project. This is usually handled by the project team working to introduce improvements with each repetition of the work sequence. Where work on a project by a subcontractor takes on a repetitive form, such as going from one floor to another on a high-rise building, then the project manager can start to look at process control techniques developed for manufacturing. This is the classic opportunity for the measurement of capability.

Assessing capability of a combination

But it gets even more difficult to measure capability; it is quite possible that there will be a combination of the possible situations. For example, one might be in a situation where it is important to get a repeating process right from the first run, and still be able to introduce improvements as one goes along.

By way of example, in an organisational change program, it might be regarded as an element of success that key members of the organisation continue to support the change. One of the activities necessary (among the many) to maintain that support might be to provide these key people with personal updates on progress. The updates will have to occur at regular periods, be good right from the start of the project sequence, and contain particular forms of information.

It is necessary to examine what form of information the production infrastructure is capable of delivering and how regularly it can deliver. It might be that the production infrastructure is capable of delivering a much higher level of service than required, in which case the capability exceeds the requirements of the threshold of complaint. It might be otherwise; it might be that the production infrastructure can only deliver reports half as regularly as required and with only three-quarters of the data. In this case, the capability is below the requirement—disappointment is almost certain. Now adjustments need to be made to get it right the first time. As work proceeds, one can streamline the process, in fact, make the production infrastructure more capable.

Assessing capability of one-off cases

In assessing the capability of one-offs, one is forced to mix the issue of *expected* output with that of the *required* output. In defining the idea of project, project sequence, and phase capability above, the word 'reliability' was introduced. In effect, we are forced to nominate the expected outcome first and then make a statement as to our level of confidence that the expected outcome can be achieved. This is a shift from the concept of capability as used in the quality literature; note that the definition given later which can be applied to processes is different from this. In stable processes, the output is predictable and is the capability—this may or may not overlap with the specifications. However, we wish to maintain that the idea of capability arises from the sequence chosen, and not from the outputs nominated as required.

So, in assessing the capability of a one-off, one identifies the objective, then identifies a sequence, and then checks to see if the sequence can deliver what is expected. We develop, at best, a feel for the reliability of the sequence in achieving the objective. This is not without its problems, which can be brought into focus by the following question: is it more capable of doing something less accurately but more reliably than it is capable of doing something more accurately but less reliably? Our concern in this chapter is to increase reliability.

Assessing capability will happen within a sequence similar to the following. This is a very similar sequence to the one for the assessment and development of the production infrastructure described in chapter 6 on the pre-implementation phase. In assessing the project sequence capability, use can be made of the following eight steps.

- identify the objectives
- describe the project sequence required, identify gross aspects of the production infrastructure
- allocate responsibilities and fill in obvious detail
- seek logical gaps and fill them[67]
- seek points consuming excess resources and reduce resources
- seek points of possible overload and resource them
- balance the activities and resources; and
- decide how to verify that the action and outcome have come to pass.

These steps are applied to each phase in turn and then to the overall project sequence. Note that this will only give a feel or a sense of the capability; as already noted, precision is not possible.

Enhancing project sequence capability

The project life cycle is a guided sequence leading to a project as output. All of the discussion in the earlier chapters was concerned to develop and manage a project sequence that would have a high reliability in terms of producing a project that would satisfy the objectives of the client. Achieving high project sequence capability is the key skill of the project manager. The following comments are brought forward to emphasise some of what has already been said, and to put it all within a context of improving project sequence capability.

1. Ensuring that the project sequence takes account of the environment.

This requires constant monitoring and checking of the market, the political environment, etc, and has been emphasised in other parts of this book. Note that some of the key resources required by the project sequence are to be obtained in the external environment. Questions such as 'What are the resources available?', 'How are they to be chosen?', 'Are the needed skills available?', 'How busy are the suppliers?', 'Have the suppliers sufficient resources to do the work?', 'How are the external relations to be managed?' and 'Has the project adequate external support?' need answers...and then further answers.

2. Maintaining clarity of objectives

It is crucial that the correct objectives are clearly identified and managed.

3. Ensuring that the whole project sequence has been identified

All of the work of the project sequence needs to be identified as early as possible. One needs to be on the lookout for missed work, often arising from a change in the environment. The work of handover needs to be planned early in the project life cycle so as to increase the chances of a successful transfer of the project to the users. By 'all of the work' is meant all of the work that will form part of the project and lead to the project, including all work of an infrastructure nature.

4. Checking viabilities

Check the financial plan, the marketing plan, the technology plan, the quality plan, the time plan, the cost plan etc.

5. Breaking the project up into manageable parts

This is a fundamental of management work. In principle, it is easy, but in practice it is quite difficult to break up work, to recognise the parts, and be able to operationalise the project sequence. The project life cycle is an important step in breaking up the sequence into manageable parts.

6. Identifying and concentrating on the key controlling items

Most project sequences will have one or more items that are critical or in a way controlling the whole sequence. They may be there because of timing (Olympic Games), cost (client has limited money), long lead items, technology, and so on. The planning work has to identify and address these items, and they may need constant monitoring.

7. Organising clear responsibilities

It is most important to identify who will be responsible for what. One has to identify the persons or organisations (or what type of people and organisations) who will carry the main responsibility for each activity in the project sequence and later actually choose them.

Responsibility matrix

One of the tools used in project management to keep track of responsibility is a 'responsibility matrix'. This is a rectangular array; against one vertical side is a list of the project participants and across the top is a list of the main activity areas. Various symbols, indicating various levels of responsibility, are then entered onto the matrix so one can see at a glance who has what degree of responsibility in relation to an activity. For example, one participant might have the final say, one might have the right to be informed, one might have to append an opinion, etc. The responsibility matrix is a useful device and a useful discipline.

Project start-up meeting

This has been discussed in other places and can be used to establish responsibilities.[68]

8. From one phase to the next, identifying the downstream customer

The project sequence has been described from the transition phase to the handover phase. To get the work done, work has to be taken from one phase and passed on to the next. The responsibility for this passing on of the work from one phase to the next is in the hands of the project manager.[69]

On many projects, the project managers change between one phase and the next. There are some good reasons for this; among them is the possibility that different phases work better under different personality types. But there is a problem in that there is a handover from one project manager to another—also, they may or may not actually meet.

Even though the project manager may not change, there is a vast change in the project sequence participants. While some participants from the earlier phase stay in place or remain available, there is a new influx of people when building up the activity of the next phase. These people bring new expectations and may have a quite different perspective on the project.

Clearly, there are potentially quite big gaps to be filled in the stepping from one phase to the next.

Where there can be personal handovers, all the better. But these are not always possible and usually leave quite a few questions to be answered. To cross the gap effectively, one needs good documentation. Having divided the project sequence into phases, the project manager needs to think about the handover to the following phase. The project manager needs to define the documentation and its contents. The documentation needs to specify in positive terms what is required and should identify the criteria for acceptance. What will be missing from this, however, will be a lot of symbolic and value data which can only be passed along by personal contact.

The issue of the quality of the documentation clearly lies in the field of that quality principle that is concerned to satisfy the needs of the downstream customer. When project sequences are continuing and when the downstream customer is in place, then meeting the needs of the downstream customer is easier. But on many projects there will be breaks, delays between one phase and the next, and usually the downstream customer will not even be in place. Thus a lot of the documentation needs to be presented in a standardised manner.

A good handover from one phase to the next is an important contribution to the capability of the project sequence. It is hard to measure this contribution but it fills in a number of logical requirements, and increases the probability of success.

The main tools in crossing the gap between phases are documentation and the project start-up meeting.

9. Managing the politics

By politics is meant the gaining and maintaining of support for the allocation of project resources and the maintenance of control in the hands of the project manager. There are no easy answers here. In essence, the answers seem to be to manage out, manage up, manage across and (less importantly) manage down.[70] All of these internal processes involve talking to people. Selling and negotiation are very useful skills for this kind of work.[71, 72] While these skills are relatively easy to describe, they need considerable practice for them to work well.

Managing out means managing the important external stakeholders.

Managing up means that one's superiors in the relevant organisations are kept informed and kept enthused, and whose ongoing support is always being monitored. Managing up involves finding out what motivates the key members of upper management and keeping them informed on what is happening. It involves quite a considerable amount of talking to them and giving the impression that all is under control. When forced to come to them with a problem, one also should have a solution (remember that the process of finding a solution is seen by many as a solution).

Managing across means keeping track and taking corrective action with people and organisations on a level similar to the project manager. This may be relatively easy when they are all located nearby, but on projects with international participants, the difficulty is compounded. Managing across involves negotiating with one's peers for support—one's peers may be in the same office or in an office half a world away. It can involve horse trading, and doing and returning favours. In a co-operative environment, this process of negotiating with one's peers can be relatively easy and based on such principles as the needs of the organisation. In more difficult environments, one needs to combine the managing up with the managing across; getting expressions of support from upper management may put useful pressure on colleagues. Managing across requires considerable time and effort, and will test the patience of many project managers. It is essential for the project manager to recognise that his or her project may not be all that important to colleagues in the office, therefore the benefits they can expect to derive from co-operating must be seriously examined and considered.

Managing down means looking after the people working for the project manager.

The interests of supporters and opponents will change. It is a mistake to assume that these interests are static and therefore do not need regular management. Plans are required to set activities in place which will monitor the interests of the relevant groups and provide tactics for dealing with them. This of course involves resources.

10. Developing layers of support

The project management work needs to elaborate the objectives. The work needs to identify further layers of possible support and the means by which that support can be operationalised. It needs to identify more stakeholders and tie in their objectives to the project. Getting further support can mean setting up a specific research team, using sales tactics, lobbying processes and so on. It needs to identify sub-objectives which the project can meet and through which it can draw in further support. These sub-objectives might be prestige or profit: financiers need to find the project attractive; subcontractors must be willing to bid for work and have confidence in being paid; environmental group agendas need to be considered; labour organisations may need to be considered; and the aims of management in the client organisation taken into account. The objectives of this diverse group will not be consistent, an inconsistency that will have to be managed. Ideally, the objectives should all be perfectly consistent, but that is the ideal. So there is work to be done developing the objectives and linking them to support for the project, which may require considerable research to ensure viability, and maintaining a high degrees of consistency.

11. Finding a project champion

A warning comes from the research literature on projects—somebody must be found who is concerned for the project itself, not just for the profit. The project management work should therefore always ensure that someone is in place who is committed to the project for the project itself.

12. Managing the technology

There are many issues under this heading, but only four will be considered here.

Concurrency

This is an area of great controversy in project management and one that has engaged the best minds in project management.[73] (Concurrency is different from fast-track in that in fast-track one is using known technology, whereas the technology is not yet resolved in concurrency.) In concurrency, production starts off before all the technical problems are solved; this is done to save time. Thus one is caught in a time pressure situation with technical uncertainty and significant commitments to suppliers. There is no answer to the dilemma. In some situations it pays off, while in others it leads to gross project overruns, both in time and money. A typical situation pushing people towards concurrency is in the area of defence, where the threat assessments are changing and one is trying to stay ahead of the perceived enemy. Similar situations can easily be seen in the civil area where competing firms try to beat each other into the market.

The general advice is to avoid concurrency wherever possible. If forced into a concurrency strategy, there needs to be very sound project control procedures and quite skilful management. One also needs to develop fall-back technical solutions, ie, solutions that will work which use standard technology and will solve the technical problem, although they may not be as good as the possible solution. If one has a project with a range of unsolved technical issues, without fall-back positions, then the prospects for project success are bleak, very bleak!

Continuous improvement versus the big step

The advent of TQM and its various advocates have highlighted the differences between going for the technological leap as against gradually improving the technology.[74] It is now quite apparent that continuous improvement as against the giant leap has had quite spectacular successes. The basis of this success is that the continuous improvement process has lead to production capability platforms that have the same impact as the giant leap. For example, in car manufacture, greater accuracy in the production of a machine can lead to smaller forces on the material which can lead to smaller components which can lead to a smaller engine—which is now a new production platform. On all projects there will be the temptation to go for a few big leaps. One big leap will be dangerous enough but a few combined on the one project raises the risk profile quite substantially. The wise project manager will keep the gradual improvement at least as a fall-back, if not the main, strategy.

Staged commitment

One of the ways to deal with the uncertainties associated with new technology is to insist on staged commitments. This helps to reduce the financial exposure and probably forces more realistic evaluations.

Reduce new technology

This reduces the risks associated with the use of unproven technology.

13. Managing a quality assurance system

Because TQM is just too difficult for some and is, in some cases, uneconomical, the general practice has turned to quality assurance. It appears that success in modern quality management is more subtle than at first realised. One of the problems facing modern quality management is the way the paperchase has taken over the quality management agenda.[75] This is regrettable. A weakness with quality assurance is that one is inspecting after the event and not really using the modern quality ideas to achieve better platforms of production.

Quality assurance requires ways of assuring people that what is being delivered is what they expect. It involves considerable levels of testing and inspection. On projects there will need to be a planned program of tests and inspections because much of the work is covered up as the project proceeds and assurance is not available as a last-step method.

The project management work will have to develop a quality assurance system.

Ensure decisions are 'traceable'

A key principle of quality management and a very important activity in project management is the maintenance of traceability. In the movement along the phases, the project manager needs to be always concerned that the path of the project sequence is always visible and can be checked and re-examined.

Project
Sequence
Capability

An advantage of this is that it places emphasis on and protects the idea of 'constancy of purpose'. There is nothing worse in project management than having to simply 'throw it all up in the air' and start again. This is a recipe for loss of control and failure, and it often happens.

14. Deliberately setting out to manage risk

Risk management has gained a lot of attention in recent years.[76] Risk management is seen as a managed process whereby risks are identified, then assessed and then responses developed and implemented. The responses may include taking action to reduce the risks that have been identified, taking out insurance, putting in some contingency to cope with the risk, or a combination of all three. The planning and design phase itself is a risk area but it also provides considerable relevant data for risk control.

Before accepting risk management at face value, it is worth noting the alternative and parallel mind-set. The alternative does not supersede risk management but should be a step in the process before the commonly recognised risk management process starts.

There are two ways to view risk management. The first is the standard risk management approach just described above—this is the common, usual method of risk management and pervades the thinking in a whole range of areas including occupational health and safety. The problem with this approach is that new ways of things going wrong are always turning up; there is an endless supply of risk.

The other mental framework asks 'What is necessary to do the job without risk?'. The framework then requires one to define in positive terms what must be done to do something without risk. This framework sometimes allows new, safer ways to be found. In many cases it will lead to solutions where risk is actually removed from situations.

Specifications to allocate risk

During the phases, risk will be allocated between the project participants. This may be done explicitly or implicitly. For instance, performance specifications allocate risk away from the designers to the suppliers. Specifications are used as the vehicle to transfer risks. There are even specifications in the construction industry where attempts are made to hold the construction contractor responsible for the mistakes of consultants.

The modern approach to risk suggests that risk be allocated rationally. It is irrational to allocate risk to people and organisations who cannot bear the risk; they may have the risk but they cannot meet their obligations should risk arise. In most cases, there is a price to the allocation of risk. Those taking on risk should cover themselves for that risk and incorporate the risk in their prices.

The project manager will have to seek advice on how risk has been allocated and then make recommendations if changes are needed.

Specialists in risk reduction
A key strategy to reduce risk is to employ specialists. The specialists know how to deal with their own risks and make it their business to be able to do so.

Warning
The mathematics of risk is based on large populations with probabilities of occurrences. As a result of this, one finds a calculation of 'expected value of risk' and this is presented as if any individual could apply this calculation to himself, herself or itself. For example, the risks might be in three groups, say $1.0 million, $0.5 million and $0.1 million. Associated with these risks might be the following probabilities (in the same order) of 0.0001, 0.001, 0.05. The expected value of this risk is then calculated as $1.0 m x 0.0001 + $0.5 m x 0.001 + $0.1 m x 0.05 = $100 + $500 + $5000 (this is in line with experience that shows that the smaller risks cost the most, they happen so often) = $5600.00. What is almost certain is that the risk figure will never be $5600.00. What is now suggested (quite wrongly) is that a project should not take up insurance if its cost exceeds $5600.00. For a large insurance company with lots of risks and clients, the outcome can average close to $5600.00 per client and as long as its premium exceeds this (plus the insurance company's administration costs) the insurance company will be ahead. But the individual faces (assuming, at most, one bad event) either nothing, or $100,000.00 or $500,000.00 or $1000,000.00. These are quite different outcomes from $5600.00. The danger here is that people confuse what applies to a large population with what happens to an individual. What is the maximum premium you would accept ?

15. Developing the project management information system

Central to the management of the project sequence is the project management information system. It needs to be developed and grown with the project sequence. Setting up the project management information system is a skilled project in itself. The system must have structure so that the data can be found when wanted; it has to be able to deal with quite imprecise data as well as data that is not quite correct but sufficiently correct for some decision-making. It is useful to separate out data that is essential for historical and audit purposes from that which is needed for ongoing decision-making. Consideration of how the data will be used or transformed into information will signal some other requirements.

The following points might be relevant when developing a project management information system:

- All data has a source and this should give a clue as to the most appropriate basis on which to build the understanding of the information flow. For instance, accounting systems are developed on the basis that a transaction has occurred—the transaction becomes the starting point.
- It is growing and changing all the time. The project management information system therefore has to be able to start off small and then grow and expand to cope with the development of the project sequence.
- Establish document trees.
- The range of documents is enormous—it includes letters, invoices, purchase orders, cheques, drawings, faxes, memos, notes, disks, and so on. The physical dimensions as well as the type of document needs to be considered; they will not all fit into a single filing cabinet.
- Unfortunately, documents rarely relate to only one item on the project or to one system. Most documents will have an impact on at least two areas. Thus copies need to made of documents and filed in the relevant files. It is amazing how quickly one can get on top of an issue if the relevant documents are all together in one file.
- There is another problem in relation to documents and that is identifying the current relevant document. Because projects are emerging as the project sequence proceeds, many documents need to be amended as the project proceeds. In the implementation phase, the correct document has to be identified otherwise work or actions will take place on a false basis.
- The standing of documents has to be identified. Some authorise payment, others authorise that action must be taken, etc. There needs to be a register of documents, regularly updated, naming the latest authorised document.
- The project management information system will need to have systems which bring people's attention to take action when that action is required. Sophisticated cash managers have systems which allow them to confidently delay payments until the last day possible.
- There is documentation that governs the relationship with outsiders (with, say, contractors and suppliers, purchase order issue) and there is documentation that governs the internal control of

operations (WBS, planning documents, monitoring, etc). The quality and control of external documents may need to be higher than for internal documents.

- The system should suggest who needs to be informed.
- The system needs to be able to collect subjective as well as objective data and recognise the status difference between them.
- Security of data is now a critical issue. This means security from inappropriate access and security against corruption of the data.
- Legal implications of holding data might need consideration.
- The project manager should develop reporting schedules in the preparation of the project management information system.
- A very difficult, but important, issue is the need for the project management information system to be able to get to bad news as well as the good news. The system needs to be able to identify suppressed data. This is an example where quite informal systems may be very effective.

This is a very difficult idea to measure but the project manager should be thinking about how reliable or effective the project management information system is. The project management information system, coupled with the quality of communication on the project sequence, is the key management weapon.

16. Using the cost system to monitor time

Cost estimates are often more accurate indicators of time requirements. In most industries there are specialists who do time planning and who do cost estimating. On most projects, two estimates are being prepared—the time estimate and the cost estimate. They are prepared by different people, one group estimates time, the other cost. But these people rarely come together. This is the astonishing aspect of project management practice. There are project planners who live for the preparation of their network and their critical path, and there are those who live for their cost records and for their analyses of tender records, but they live in separate worlds. While these two estimating specialists have different sources of data, they are actually either directly or indirectly measuring productivity.

Turning now to the cost estimating people, they identify products and work to be done. They then gather prices for this work in the past. They do not generally estimate time. The key to their accuracy is their classification of work so that it can be compared from project to project. So the production of a piece of software can be broken down into various work components, a component such that its size varies from job to job, not its nature. The cost estimators then develop rates for the work. Their final estimate is then the summation of the rates per unit of work by the amounts of work plus other allowances.

What is interesting about these cost estimates is that they must have incorporated in them the equivalent of industry average expectations of time required, that is, of time productivity. The time is not necessarily explicit. The lack of explicit time does not deny the time data, it is in there but hidden. Therefore there must be a correlation between the cost estimates and the time estimates. The cost estimates contain within themselves implied estimates of task duration.

It seems surprising that in many competitive project businesses, one is faced with claims of tight financial margins, but there are huge time overruns—yet companies survive. It is very odd that a competitive business like building construction where margins are seen as slim and tight, where people talk of margins in the order of two to four per cent, that companies can spend 20 per cent or 30 per cent more time than expected on jobs and still not go bankrupt. Somewhere an allowance has been made for the time overrun. This is not in the explicit margin which is low so it must be in the rates themselves. *One explanation might be that the cost estimation system has got a better time estimation capacity than the standard time estimation techniques!*

Time estimation techniques are derived from one source of data; cost estimates are derived from another set of data. Both contain data on task duration—in one, the data is explicit; in the other, the data is implicit. The results from both processes should be compatible. Major discrepancies imply problems. Such discrepancies on projects are rarely examined; it is only a small number of companies that take the trouble of checking the compatibility of the data.

17. Taking advantage of processes

Emphasis has been placed on the one-off nature of projects. But it also has been said that a project manager should set out to reduce the amount of one-offs in the project sequence and thereby reduce the risk to the project. Identifying and establishing repeat steps, or identifying and establishing processes, represents a major step in developing project sequence capability.

When the processes have been identified and established, there is a whole range of experience in manufacturing which can be incorporated into the project management sequence. Most of this will be applied at subcontractor level and below. It can also be applied to many of the project management information system activities, which usually involve processes rather than one-offs.

The following are some introductory comments on quality management. The reader is referred to texts for more detail.[77]

Control of the process

Getting the process to produce what is wanted involves stabilising it and then moving it to the standard required. One stabilises the process by removing the *special cause* of variation.

A *special cause* of variation is an assignable cause of variation. It is specific to some group of workers, to a specific machine or to a special local condition. It could be the weather, different fuel or a different worker.

A *statistical chart* can be used to detect the existence of a variation due to special causes. Note, however, that it does not find the cause. On a statistical chart, a special cause is indicated by, among other things:

- seven successive values below average
- a trend of six consecutive points.

Removing all the special causes leaves a stable process.

A *stable process* is a process with no indication of special causes of variation. A stable process is said to be in *statistical control.* Variation is only due to random causes. As long as a process is stable, the distribution of its output is stable and predictable.

One then has a process in statistical control. The only causes of variation are now *common causes.*

A *common cause* of variation is a random cause of variation in a process that is in statistical control. If only common causes of variation are present, the output from the process forms a distribution that is stable over time and is predictable. If special causes of variation are present, the process output is not stable over time and is not predictable.

Going back to what was mentioned above in relation to distinguishing special and common causes, two kinds of mistake are possible.

Mistake type 1 is calling a mistake or variation a special cause when in fact it is a common cause.

Mistake type 2 is calling a mistake or a variation a common cause when in fact it is a special cause.

One improves the process by changing it so as to reduce the common causes. The work of improving the process by removal of special causes and then developing a new process is the work of the specialist disciplines. The management approach is that those in or close to the process can deal with special causes but only those with a more overall responsibility for the whole system can deal with the issue of common causes. This has a key impact on how one manages improvements, and the division of responsibility.

A system or process that is in statistical control has a definable output known as the *process capability.*

The full implementation of these ideas on a project requires a large project. However, they are quite applicable to subcontractors who do similar work on project after project.

A statistical note

The developers of statistical methods of process control have introduced very interesting simplifications to the statistics. Usually there is the issue of what kind of distribution is involved; deciding this requires considerable statistical skill. The developers of process control charts have avoided this issue by plotting key properties of sets of samples rather than the properties of single samples. This move allows them to stay with the normal distribution. This is based on the Central Limit Theorem.[78]

This means, in simple terms, that while the population may have a distribution other than normal, if one takes samples from the population and finds the average values of the samples, the distribution of the averages will be closer to the normal curve than the original population. Hence in the use of charts, each sample consists of a number of pieces. This is the basis on which it is possible to conduct a vast range of statistical activity within the framework of the normal distribution.

18. Strengthening project sequence capability: other issues

The following questions are pointers to other areas to be addressed in strengthening project sequence capability.

- How is the project to be funded?
- Is the funding flexible enough?
- How secure is the budget?
- How long will it take to finish the project?
- When will design be complete?
- How will occupational health and safety be managed?
- How will industrial relations be managed?
- Are the specifications adequate?
- What infrastructure is required and how will it be obtained?
- When must commitments be made?

Quality and sequences

Sequences provide a special challenge to project management and a great deal of work needs to be done. The development of project management requires the development of ways and means of measuring and improving the project sequence capability.

Part Three

Keeping it all together

Chapter Ten

10

Project Communication

Project communication has a dual purpose: the first is to cause some action or agreement to take place, and the second is to make a record which might be needed later (perhaps in case of argument such as a disputed payment).

A word of warning to the introverted, strongly technically orientated project manager—the technology is only part of the project; getting it to happen requires vast quantities of communication, usually in the form of lots and lots of talking.

Interpersonal communication

Project managers need interpersonal skills. Interpersonal skills involve dealing with people, facing conflicts, persuading people and much more. They also mean being able to behave in a manner appropriate to the situation and to the person or persons involved. The key to interpersonal skills is flexibility of behaviour.

Behaviour, then, has to be modified depending on the situation; different situations require different behaviours. Project managers must therefore develop behavioural skills in order to give them the flexibility they will need. This is a tall order, but it should be the objective of behavioural training offered to project managers.

One should have great reservations about behavioural training conducted within the context of an ongoing work-group. Such training should be conducted away from the people the project manager will have to work with later. When learning about behaviour, there is the need to be able to experiment, for example, giving vent to feelings about a work-mate, without running the risk of jeopardising relationships.

The author recommends that people interested in this area might follow up training in the form of T-group training and in the form of Neuro-Linguistic Programming (NLP) training.[79, 80] This is best done away from work colleagues in a safe environment under the guidance of a skilled and experienced practitioner.

Developing a framework to guide behaviour

There are many texts and courses on methods of interpersonal behaviour. Readers wishing to read further in the area may refer to these, and also to texts on negotiation and selling.[81, 82]

But before the project manager goes into methods and techniques, there is a much more difficult issue to deal with. That is the framework of values the project manager will use as a guide to communicating. Many of the project manager's activities take place in a political context, a context in which power and values come up against each other. In such situations, the project manager's fundamental values are tested. The practicing project manager needs to learn from these experiences and then give consideration to the values required. But remember, it is very difficult to alter fundamental values.

Establishing openness and trust

In project management education, there seems to be almost universal agreement that interpersonal skills are vital for project managers. Much of the literature seems to equate openness and trust with interpersonal skills. It is doubtful if this is correct. Openness and trust are skills needed in some forms of interpersonal work, eg, counselling, but they are not the only ones, and may not be ideal in project management.

Honest, professional behaviour is recommended, but this does not necessarily imply openness and trust. Being honest is not the same as being open; being professional is not the same as being trusting.

There is no doubt that the style in which the project is managed can develop openness and trust. Openness and trust can lead to great effectiveness and efficiency—there will be times and situations where these qualities are vital for project success, or for the general efficiency of the project. In such cases, openness and trust need to be established and managed in a very professional way.

Wherever possible, the project manager should have things so arranged that people do not have to rely on openness and trust; everything of importance must be independently testable.[83] Further, the project manager must avoid situations in which conflicts of interest may arise, or, if such situations are unavoidable, develop clear processes, independent of the will of participants, by which interests are managed.

Using language and NLP

One hears much about the appropriate use of language. Neuro-Linguistic Programming (NLP) is a technique that provides specific advice on language use, and it is recommended that the reader looks at some texts on it. NLP shows how language can be adjusted to suit the needs of different people.[84]

The following is a fairly unsophisticated example. Something has happened, something has been done. We are not discussing a painting, a piece of music or the tactile features of a piece of material, but the fact that someone working on the project has done a good job. Three comments could be made:

- It looks good
- It feels good
- It sounds good

The question now is which comment does the reader prefer? Most readers have a clear preference for one—architecture students prefer the 'looks' comment, while most building students prefer the 'feels' comment.

Reports written in short, clear sentences suit clients who prefer the 'looks' comment, whereas reports written in long, well-constructed but well-qualified statements are preferred by clients who like the 'sounds' comment.

NLP has a lot more to offer in the field of interpersonal communication than this, and the project manager might find it useful as an underpinning to negotiation and selling.

Negotiation

Negotiation is one of a project manager's most important tools. It can be combined with personal selling skills, discussed below, to form an important set of communication skills. There are some very good books on negotiation; the book entitled *Getting to Yes* has achieved the status of a classic, and is highly recommended.[85]

A major advantage of having an approach such as that outlined in *Getting to Yes* is that technically orientated project managers have a framework within which to operate in the more subtle areas of social interaction. But reading the text is not enough.

The book puts forward a number of principles to guide the conduct of negotiations. Although the principles are easy enough to understand, to bring one's behaviour into line requires considerable skill and discipline. It is amazingly difficult to shift one's mind-set away from positions to interests, and also to be able to separate out the people from one's emotions. This in no way detracts from the book; just note that practice is needed, particularly in separating interests from positions, and in getting emotions under control.

A really useful insight is BATNA (referred to earlier), the Best Alternative to a Negotiated Agreement. This puts the bottom line in a completely new perspective and is a very important point for a project manager to identify before entering negotiations.

It is, however, recommended that project managers also look at other publications in this field.[86] They often cover other aspects and give advice on specific tactics and rules, some of which are not explicitly stated in *Getting to Yes*. For instance, the rule that once you have made a proposal, the other side must make its own proposal and not simply sit there firing arrows at you, is an important tactical issue not covered in *Getting to Yes*. (It is acknowledged that *Getting to Yes* deals with such tactics in a different way.)

The practice of negotiation is quite interesting to observe. Much careful discussion, with people going to and fro takes place, and one thinks it will never end. Then suddenly everything is agreed—the last stage of negotiation can be breathtakingly quick.

It is strongly recommended that the project manager reads some books on the art of negotiation, particularly *Getting to Yes*, and starts practising how to separate interests from positions, how to control emotions, and to develop a good understanding of BATNA.

Personal selling

For all its unsavoury connotations, selling is an important aspect of a project manager's activities—ideas must be sold, stakeholders must be persuaded to continue to support the project, and an attractive view of the project must be presented at all times. There are people for whom selling is almost a natural activity, but for most, it requires some learning.

There is a logic to selling; it is not a random process. It can be planned for and the chances of success increased. The logic of the selling methodology can be applied to situations beyond selling. Many aspects of selling can be learnt, but the learning needs to be tailored to the personal value system and style of the project manager.[87]

Selling the benefits

Many selling methods exist. However, one that project managers will probably find acceptable is the method often referred to as benefits selling. Here, the benefits of the product or service are matched with the needs or wants of the prospective buyer. The method emphasises benefits rather than features; it essentially 'qualifies' the prospect and then moves them along a decision-making process, mainly by asking questions. The process recognises that people do not make buying decisions out of the blue but arrive at a decision as the result of a process. One of the purposes of the questioning is to identify how the prospect likes to be dealt with, and then to deal with them accordingly. The benefits selling process is very focused on the buyer.

Qualifying the prospect

This method involves targeting those people who are—in the jargon—ready, willing, and able to buy. A qualified prospect is one who has the authority to buy, has the wherewithal to buy, has a need or want for the product, and has a timeframe within which it is intended to complete the purchase. So effort is, first of all, directed towards finding a prospect and then seeing if they satisfy the qualifying tests. Unless the prospect is seen to be qualified, or can be persuaded to qualify, there is little point in continuing to try to sell to them. If the project manager is trying to sell an idea or an approach, then someone who can use it must be identified, someone who can get the authority to act on it, someone who can resource it, and someone who also has a timeframe within which to implement it. In order to qualify a prospect, one needs to gather information; this is achieved mainly by asking questions. As can be seen from this first stage of qualifying the prospect, there is a logical framework behind this selling approach, a framework that can provide guidance on what has to be done.

Buying decision process

The prospect goes through a set of steps before arriving at the decision to buy. A whole body of research in consumer behaviour exists which tries to understand these steps.

The steps consist of the prospect considering the facts presented by the seller, facts that are always linked to benefits; the seller then handles objections and finally closes. A difficult skill to develop is the linking of facts to benefits. Facts on their own are actually quite boring; they only really come to life when put into some meaningful context. For the technically minded project manager, this requirement is quite difficult to master.

What is very important is the link to the emotional benefit. If no emotional benefit can be found, the facts and the other benefits fade in importance, becoming practically irrelevant.

Handling objections

The next interesting part is the handling of objections; the selling literature is full of routines for this. Many of the techniques rely on questions with prospects often handling objections themselves. One has to distinguish between real objections and pretend ones. There are neat ways of doing this, which a project manager could find useful in identifying core problems and not wasting time on irrelevant issues.

Convincing people

Persuading people to buy can be interesting and, in many ways, quite surprising. Most project managers might imagine that a buying decision is made on the basis of careful study. In fact, many people have very simple ways of making a decision; the statement, 'If he likes it, I'll have one' is often heard.

Closing

This is where the prospect is asked for a decision. In many situations, the decision the seller is looking for is agreement to meet again, not the decision to buy. Later, when the decision-making process has been allowed to take its course, the 'buy' question is asked. This is a useful process for the project manager. Time and time again, the project manager has to ask the client for a decision of one kind or another. After asking for the decision, the project manager should remain silent and let the client consider.

When the prospect agrees to buy the product, the seller briefly confirms that a good decision has been made and moves on to minor procedural matters—one does not revisit the decision, discuss it or chew it over. There was one trade union organiser who, as soon as he got a decision, left the meeting so as to avoid the possibility that the decision might be reversed.

Questioning prospects

Open form of questions are recommended, except when asking for the affirmative answer that closes the sale. Open questions usually start with one or other of the following words: what, why, when, where, how and who.

Oral communication

A brief comment on the use of oral communication on projects: first of all, there is an awful lot of it—in fact, it is the key form of communication. Written documents often serve to simply record what has already been agreed during discussions.

Oral communication is rarely recognised as part of a project's management information system, which is unfortunate. A lot of project problem-solving takes place without there being a written record of it. The power of the oral system does not seem to be well-understood. In any managerial hierarchy, there appears to be a line above which work is done through oral communication as well as through significant amounts of written records. Below this level, the main system of communication seems to be oral.

Down to a certain level, usually just above the level of the lowest-level operatives, much oral communication is interactively solving problems. At the operative level, most oral communication appears to be personal and social. One reading of this is that different attitudes towards work prevail; however, a much more compelling functional explanation is available. Once told what to do, operatives have skills and can do it. If they encounter problems, they don't need extensive interaction to solve them; their own skills solve them. As one moves up the hierarchy, the problems begin to interact across various boundaries and people higher up need to discuss solutions.

At the project manager level, there will be many problems requiring interactive problem-solving (as noted earlier, a strong force in favour of flatter hierarchies). Although this is exhausting, the project manager's importance and existence gets considerable confirmation.

Inter- and intra-team communication

Very few people see a project through from beginning to end. It is, in fact, rare for a project manager to see the whole project through, and if one does, it is most likely to be the client project manager. Other people come onto the project, stay for a while, and then leave. It is quite common for many people, particularly those engaged on the implementation phase of a project, or those working on defence projects, not to know the nature of the project. Throughout all this extraordinary turnover of people in a project sequence, a sense of coherence must be maintained.

There must be a source of stability. It is important that the client organisation is seen to be stable and its leadership passed on in a very controlled way. Passing on the leadership can be a dangerous time for the project and, where possible, the incoming project manager should be at work before the outgoing one has left.

On the supply side, the number of people involved builds up in increasing waves, then reduces towards the end. The feasibility phase usually sees the hiring of experts across a wide range of important disciplines; usually, these people work together as a team. In the planning and design phase, many more people work on the project sequence, but may in fact be quite removed from the project managers. In this phase, a network of organisations contributes to the project. The number of people involved explodes in the implementation phase and the network dramatically increases. The larger projects involve hundreds of networked organisations. In the handover phase, numbers are reduced; a new team may come in to manage this part of the sequence.

There are two parts to the project manager's role in dealing with organisations on projects. The first part is the project manager's team, which could be made up of a number of individuals working separately or who could possibly become a very effective group. Beyond this is the second part, which is the management of a growing, then stable, then ultimately declining network of other organisations.

Managing a team is probably helped by some knowledge of group dynamics; this subject is discussed later on in this chapter on p.236.

The networked organisation is usually brought together by contractual arrangements and is managed by a process of interface management. Part of managing the interfaces involves the use of interpersonal skills and meetings. On some projects, but not many, the participants in some of these meetings become cohesive groups. On projects where long involvement is required, say over six months, it is probably worthwhile for the project manager to try to develop a degree of cohesion between those who participate in the more important meetings. In doing so, however, it is important that 'group-think' be avoided.

In summary, the setting up and management of temporary organisations is the project manager's responsibility. Building temporary organisations involves getting the physical facilities together, finding and appointing staff, finding and appointing subcontractors, developing teams and allocating tasks; it also often involves managing conflict.[88]

Setting up a networked organisation

The usual form of organisation executing the project sequence consists of a central client team linked to some individual supplier organisations and to one or more networked supplier organisations. The individual suppliers directly linked to the client are usually specialist consultants, often in some form of professional relationship with the client. There is usually only one networked supplier—the main supplier linked to a range of subsuppliers, or subcontractors.

The sequence of building up and dismantling the temporary organisation for the project therefore involves the gradual build-up of a team or functional hierarchy to run the sequence and, as this is going on, connecting to and supervising a range of suppliers.

The organisational model where there is a central element which connects to and then lets go of other elements as required is very flexible.

It is the model that is putting pressure on larger organisations to become smaller and to outsource their work.

The setting-up of the temporary organisation for the project sequence is a divided responsibility between the client project manager and the main supplier project manager. While this, in practical terms, is a shared responsibility, in practice it is legally separated. Each has a vested interest in the functioning of the various parts of the project organisation. The client project manager is concerned that the supplier network functions well and the supplier project manager is concerned that the client organisation plays its role effectively.

Basically, there is a client organisation linked to a supplier organisation which is networked to a range of suppliers. In various industries influenced by project management over a period of time, there are often long-standing relationships between the main supplier and the networked subsuppliers. Essentially, the client is often choosing between one network and another, perhaps without being aware of them. Often the client thinks only of the main supplier, and not the attached network.

The client team may be large enough to need a functional hierarchy, but this is rare. The main supplier team, which is linked to the client and also linked to a range of suppliers, very often extends into a functional hierarchy, usually quite shallow. The supplier project manager requires a project team, or functional hierarchy, to manage responsibilities which for the most part consist of the appointment, supervision and separation of the various subsupplier organisations.

When each supplier organisation is appointed, it must also start to build up a team of people working towards the project, and then later downsize it. Each supplier often has the equivalent of a project manager who is charged with delivering the supplier's work on the project sequence.

In the networked organisation, the functional relationships and the contractual relationships do not have to coincide. This often adds to the project manager's problem of having responsibility without authority.

Managing the networked organisation

This requires special skills. The basic problem in managing the networked organisations is that the project manager does not have contractual power to direct the suppliers in how they are to do their work but does have responsibility for the effect their output has on the project. The supplier cannot just be left to get on with it on the assumption that at the end everything will be satisfactory. The project manager must get involved, but involved in an influencing way rather than a directing way.

Clearly, a lot of interfaces have to be managed. The best thing the project manager can do is to stay well-informed on progress and have sufficient milestones in the contract to allow some pressure to be exerted on the supplier. Project meetings are important in building and maintaining that pressure.

Where there are no contractual links, the client project manager has to do a lot of managing up and managing across to ensure that support for the project sequence is maintained. It is a skilled political role. The project manager must have supporters in key places and on key committees.

Internal management

Within each of the networked organisations, there needs to be some internal management. Much of this management focuses on running teams and groups. The importance of teams is an issue in project management. Teams and project management are discussed in the next section in the context of group dynamics.

Managing groups and group dynamics

Group dynamics is the area of study concerned with how people behave in small groups. Groups and teams are important; much of the work done on projects will be carried out by groups or teams. The distinction between groups and teams is a little arbitrary. A team is a collection of people who do not necessarily see themselves as working closely together, for example, a management team where the team is simply comprised of the managers at a particular level in an organisation. A group is a collection of people who have developed some sense of identity and cohesion. This distinction, however, is of little use, and the concept of a team will be incorporated with that of a group.[89]

When small groups of people come together periodically and interact, over time the nature of the interactions changes and develops. These changes can be for good or bad. Some people, probably very many, believe that it is possible to guide these changes so they will be for (what they perceive to be) the good.

In the area of group dynamics, a range of ideas about the behaviour of groups has developed. The most significant concepts are probably to be found in the terms, 'group cohesion', 'role theory', 'phases of development of the group', 'norms', 'group-think', 'task leadership' and 'emotional leadership'. The theories underlying these concepts provide insights into the behaviour of groups; from such insights project managers may be able to find guidance on how to act in a particular situation.

Groups form when people come together in such a way that they can interact. Processes occur that seem to set up an entity, ie, the group, independent of any one member of the group. The group takes on a life of its own, though it is a life dependent on the people in it. In fact, people do things in groups that they would not do alone. For the group to continue to exist, its members must interact. Most project teams become groups with varying degrees of cohesion.

Group cohesiveness

This refers to the degree to which members of the group are attracted to each other and are motivated to remain in the group. The degree of cohesion can be measured by psychological testing, but any moderately sensitive project manager should know a cohesive group just by the 'vibes' it sends out. Highly cohesive groups can greatly influence group members, are more effective in achieving goals that the group members have themselves set, have higher member satisfaction and survive longer.

As regards productivity, it appears that while highly cohesive groups are not necessarily more productive, they are more consistent. This is a judgement the project manager will need to make when considering the extent to which cohesive groups are desirable or should be developed.

As a starting point, it is probably best to aim for high cohesion and to attempt higher production levels as a group effort. Should the group fail or plateau out at a lower production level than that required, the group can be broken up. Cohesion dies very quickly if not maintained.

Norms

Norms are the unwritten rules that people are expected to follow. Members of the group expect each other to follow the rules and to hold by certain beliefs. Norms help people to identify with the group, to know that they belong, and are sometimes backed up by rituals, such as eating together. The development of norms takes place unconsciously and automatically. They are not fixed for all time, but changing them causes considerable emotional upheaval in the group. Any sensitive person can develop an awareness of the group process and know what the norms are and when they are changing. T-group training develops this skill.[90]

The danger is 'group-think'. There is strong though unconscious pressure to conform to a group view. This might be good in maintaining morale but it is dangerous when the group has to solve problems, particularly when answers cannot be found or it is uncertain what the next step should be. A project manager should attempt to ensure that somehow or other critical questioning survives in the group.

The process of group development

As time goes on, the group develops and changes. There are various models of the process of group change but a useful one is the idea that a group goes through a set of stages in becoming a more cohesive group. The group starts at one level and seems to stay there; it then goes through a phase of dissent; the dissent is resolved and then the group moves to the next level. Throughout this, cohesion is developing. Some of these stages involve conflict and are quite important in developing the functioning of the group. They usually lead to the development of the group hierarchy and the emergence of specialist roles.[91]

The project manager must recognise that the development of the group requires these times of dissent, which can get quite emotional. Any attempts at suppression only serve to suppress the development of cohesion. The project manager, however, has to be able to distinguish between dissent that has some serious threats associated with it and dissent that is part of a normal, healthy group development.

Roles

There is an area of study called role theory which can, very crudely, be likened to the definition of specialist jobs. It is just that the jobs are not so clear and easily defined. Role theory examines the idea of 'role'. A role is a way of behaving or a way of viewing behaviour, with a few complications added on. The difficulty is that there are different definitions of role.

One way of defining role is to say that it is a set of expectations (eg, norms, taboos and responsibilities). These expectations are the expectations a person, and others around that person, holds. The expectations can vary over time.

If you are a mother, then you and the rest of the world have expectations of your behaviour as a mother. Not everyone has the same view about a mother's behaviour, however, so there are different expectations of it. Living with different sets of expectations of your behaviour is very stressful. A project manager is a role; there is a set of expectations as to how the project manager should behave. Different participants in the project have different expectations, with the result that some project managers, unless they are very aware of all this, find the situation intolerable. Role theory points to factors that can be a serious source of stress and discomfort.

Defining roles can be useful because confusion in this area often leads to problems such as tasks not being done (the 'That's not my job' attitude) or being done differently and repeatedly. Other problems that arise include role ambiguity, role incompatibility, role conflict, role overload and role underload. The project manager needs to be aware of the possibility that role problems can disrupt teams. Corrective action includes making specific statements on what a role constitutes, defining boundaries and reducing workloads. Using the notion of roles can help to solve some significant team problems.

When setting up a project team, the project manager is also setting up a role system—a set of roles aimed at achieving a common objective. The roles will change and evolve over time under group pressure, and the project manager has to manage both the people and their roles.[92]

Problem-solving in groups

Groups give variable performances when problems need solving. When ideas from different skills are required to solve a problem, a group can be very effective. If fresh ideas are sought, techniques such as brainstorming may be useful. However, where the task is quite complex, the evidence suggests that the group should work to a plan to find a solution. The project manager is an appropriate person to devise the plan and monitor its progress according to the plan.[93]

Exercising leadership

This is an area of great interest in both management and project management literature. Leadership is, however, an elusive quality and can be expected to remain so. It will remain elusive because once it is known how people are made leaders, the competitive elements in leadership struggles will change the ground rules, and so back to square one.

There is evidence that people share a common perception as to who is the leader of their group. However, there is almost no agreement as to what constitutes leadership. None the less, the following points are made.

There are two quite different forms of leadership, namely group leadership and position leadership. Group leadership is leadership in unstructured or informal situations, whereas position leadership relies on the existence of a formal position. Armies down the centuries have organised things so that they did not have to rely on group leadership but on the more secure positional leadership. However, in project management, given the lack of formal authority, informal leadership skills are a great advantage.

Another topic which will be touched upon is the issue of power and its use in project management. While power is a difficult and complex concept, it is worth mentioning that the project manager should carefully evaluate the rewards that can be offered and the punishments that can be administered. These rewards and punishments may be substantial, but they may also be symbolic. The project manager must also develop skills in exercising influence, many of which derive from negotiating and selling skills. The possession of influence and the ability to deliver rewards or punishments is a 'powerful' combination.

Task and emotional leadership

An early outcome of research suggests that two leadership roles have to be played in any successful group—there is a functional role which promotes the execution of the task and a functional role which helps to manage people's emotional involvement with the group. Often these two roles are split, with one person being the task leader and another the emotional leader. It is rare for one person to be able to satisfactorily undertake both roles.

More recently, further research has increased the number of functional roles necessary for effective group functioning.[94]

What is important to note is that there are specialist roles which assist group functioning. A very important set of these is concerned with emotional issues. Emotional issues are about feeling part of the group, about being valued by the group and about being liked or disliked.

Symbolic activity

It is probably very important for the project manager, particularly one who comes from a technical background, to recognise that a significant amount of management and leadership is concerned with the manipulation of symbols.

Symbolic activity is usually quite subtle and therefore difficult to define. It can be described as that activity or behaviour that confirms or denies the order of things without actually appearing to do anything substantial. It may or may not involve verbal expression. Examples are behaviour that acknowledges or denies who is in charge (walking behind or in front of someone), behaviour which is there to save face or avoid insults, or the opposite (behaving so that no open conflict takes place, or ensuring it does), behaviour that acknowledges the existence or absence of friendship (a wink and a nod, or a cold stare), expressing a belief that all will be okay (even though there is no evidence to support such a view), saying nice things about the person leaving (and quite deliberately avoiding reference to all their less favourable aspects), the person leaving saying how wonderful it was to be part of the group (an important piece of symbolic activity which confirms the ongoing importance of the group), a selection procedure whose purpose is to legitimise the appointed person rather than find the best person, and the behaviour of the selection panel who, after a long-drawn-out brawl over the selection, agree that the recommended appointment is the 'unanimous view of the selectors', and so on. Symbolic activity is surprisingly powerful.

In management, there is a lot of symbolic activity whose purpose is to continue to assure people that all is well and under control. Even when things are clearly out of control, a project manager who appears to be calm and in charge, and who knows what is to be done, is obeyed and followed. Project managers can also create the illusion that all is well, even when there is no substantial evidence for it; the odd thing is that things do turn out well, largely because the illusion was generated and accepted. Even when things turn out wrong, stories are invented to put a positive light on things. The method of stories is a very important element in the management of symbols.[95]

Some of the more important symbolic activities are conducted within the framework of ceremonies. Everyone knows that modern buildings do not in any way rely on a foundation stone for stability, but it is often useful to have a foundation stone-laying ceremony with people applying mortar with little trowels. Such a ceremony is a public affirmation that the project has support and will be completed; it does not contribute in any physical way to the success of the project. Knowing the use of what is useful is important, but sometimes it is more important to know the use of what is useless![96]

Clear symbolic acts can be carried out, such as the way the appointment of the project manager is announced. A few lines on an obscure page of the company magazine is quite different from a front-page spread with a photograph of the new project manager being congratulated by the chief executive. The manner in which someone is named as project manager is a symbolic act and signifies quite a number of things.

Having been nominated as such, the project manager then does things that symbolise the new position. Something might be refused simply to indicate that things can be refused. Much symbolic activity would have been undertaken by the project advocate way back in the transition phase. Just because the project is under way is no reason to think that the need for symbolism has disappeared.

There are many situations or issues which are not amenable to rational processes, such as anxiety or a group's 'hope of success' on a project. There are situations where a problem is painful to deal with and something needs to be done to reduce the pain, for example, the forced exit of a group member. There are also more primitive matters to deal with, such as personal loyalty, saving face and acknowledging authority.

It is quite likely that a successful leader exhibits considerable symbolic activity. Thus project managers need to exude confidence in the project sequence, give recognition to contributions, call on everyone to think of the greater good, take part in project rituals, and so on.

This area is worthy of further study. For the moment, however, it has only been discussed in general terms. Project managers should be alerted to the fact that symbolic activity can be quite powerful and—what makes it quite difficult—it is very subtle.

Situational models

The form of leadership style that should be adopted concerns many managers. Should one be autocratic, democratic, or *laissez-faire*? The answer appears to be, 'It all depends'—it all depends on the particular situation.[97] One can be democratic in one meeting and autocratic in the next. Although people can be very flexible and can change their behaviour, they cannot usually do it in a controlled way. It is helpful if the project manager can develop the ability to change management style according to the situation; to match the situation with the appropriate style.

If the project manager is uncertain how to behave, unfortunately the autocratic style is more likely to be successful.

Communications with the general public

In the marketing area, the need to address the public is explicitly recognised. Marketing theory deals with it under the heading 'promotion', and marketing people conduct promotions as part of an ongoing selling process. It is worthwhile thinking about the ideas of promotion in terms of project management.

Promotion consists of advertising, sales, sales promotion, public relations, publicity and personal selling. Although these concepts are discrete, some overlap occurs. What the project manager has to do, sometimes with the advice of the appropriate specialists, is decide on what is called the promotional mix. This is the mix of the above concepts, advertising, sales, etc, that is to be used. In general terms, personal selling is the most expensive form of promotion in terms of cost-per-customer contact. So the main concern in reaching the decision on the promotional mix is the amount to spend on advertising and the amount to spend on personal selling. According to marketing literature, there will be a tendency to favour advertising when:

- the project's stakeholders are geographically dispersed
- the project's supporters are silent and dispersed
- there are many stakeholders
- the project is fairly standard
- the project message is non-technical
- the project has hidden qualities that need to be revealed
- there are powerful emotional reasons to support the project
- the project has enough money.

When the opposite of the above conditions applies, the decision may be in favour of personal selling.

Management of external political and community issues

Maintaining support for the project is vital in project management. This involves ensuring that the project has the continued support of the client, stakeholders and others. Inevitably this means project managers have to spend considerable amounts of time lobbying politicians, keeping them interested and watching for any waver in support. To the technically minded project manager (and probably to many other people), this seems an awful waste of time and effort. There is a project sequence to run, after all!

Those projects which have overwhelming support probably only need it monitored, but those that have both active supporters and active detractors really need attention. Such projects may require specialist management which, in this area, is provided by lobbyists and public relations companies. To the uninitiated, it can be a great surprise to see the extent to which the news is managed.

The project manager cannot simply appoint the specialists and walk away. Of great importance is the story or angle to be conveyed about the project. Some of the activity of public relations experts and lobbyists can be subjected to the control framework; control should be applied to the messages the project sequence is sending out. The skill of the specialists lies in their ability to get the message across, but the content of that message should be reasonably questioned and evaluated by the client project manager. Though difficult to do, attempts should be made to monitor how effectively the message was received.

Taking account of the external environment

There is widespread agreement that the external environment contains the major threats to a project. This recognition is relatively recent and has opened up new areas for project management attention. The project manager can no longer simply pay attention to the internal project environment.[98]

It is a project management failure to concentrate exclusively on internal issues. Issues such as time control, cost control and quality control are internal issues, because, for the most part, they are within the control of the project manager. They are often technical in nature and of great interest to many working on the project. Progress on these internal issues leads to a sense of progress on the project; they rarely if ever pose a threat to the project, or lead to complete project failure.

While threats to the project are, if anywhere, generated in the external environment, it should be remembered that so too are the opportunities. The reason for the project's inception and existence lies in the external environment, and the project manager must not overlook the significance of that.

Conceptualising the external environment

The external environment needs to be conceptualised so that project management can map out and develop appropriate responses to challenges presented by that environment. Large civil engineering projects are not the only ones affected by the external environment; it has an impact on such projects as company marketing strategies, film distribution, banking changes, the introduction of pay television, the development of pharmaceuticals and publicity campaigns. The introduction of the tax file number system by the Australian taxation office was a large public project with considerable scope for generating resentment in the community.

Identification of groups

In order to conceptualise the environment, segments of it need to be identified and classified. What is always found is that there is a political and/or social group or groups associated with each classification. The client project manager has to identify these groups and decide how to deal with them.

A vast number of groups have an interest in the physical environment. Any project manager running a project that impinges upon this environment must be keenly aware of the fact and also take into account the interest groups. Projects affecting an area of business, or part of the economy, will throw up groups representing business and organised labour. How the legal environment works is another area the project manager must understand.

While there are many ways to identify and classify the various groups in the external environment, it is essential that strategies be developed for their management.

Control and influence

Most of the external environment is outside the control or influence of the project manager—the weather, the economic climate and technology, for instance. Project management can only try to take advantage of the opportunities, or defend the project against the various threats presented by these elements. Attempting control over the external environment means that uncontrollable elements must be monitored so that appropriate adjustments or defence strategies can be put in place. The difficulty is to decide how many resources should be spent in this way.

Ideology

The project manager needs to look at the issue of ideology in order to:

- promote the project as an entity worthy of public support;
- make sense of people's behaviour in the area of industrial relations;
- ensure the project is being conducted in a way acceptable to others;
- build up motivation; and to
- persuade people to co-operate with the project sequence.

Ideology is a very powerful area.[99] It can be seen as a set of principles which people should use in helping them to decide how to behave (privatisation is a doctrine), or a story or narrative which deals with some important issues, and gives guidance on how these should be handled (Wagner's *Ring Cycle* might be saying that power and love do not go together) or it may be symbolic. Symbols represent things, often quite intangible things—a salute is a symbol, a flag is a symbol. People can feel quite strongly about symbols, as has been demonstrated by the way they respond when national flags are burned.

Winning the ideological battle

Ideology comes into focus in industrial relations but it is also very relevant when dealing with the external political environment and community issues. Projects almost invariably alter things for people. Projects bring about change which benefits some, but not others; it is rare for a project to benefit everyone. Whether or not this change is good and fair, depends upon one's values, on one's ideology. The project manager must attempt to influence the ideological agenda in favour of the project's interests, which means acting at the public and political level.

The ideological battle must be won. There needs to be a sufficient number of people, with enough power, supporting the project to ensure it gets finished. The ideological landscape is an area where right and wrong are not always clear.

Developing a project ideology

Project managers needs to develop an ideology which promotes and protects the work of the project sequence and themselves. This means that they need to be thinking of doctrinal justifications for the project (it gives employment, it is good for the environment, etc), need to be building up positive stories about the history of the project and why it exists, and need to be appropriating and developing symbols for the project. In doing so, they will be confronted by their own personal values, their own world views and moral values, and their own interests, and these will act as constraints.

In developing the ideology, project managers must think of the various groups to whom it must appeal (there can be more than one ideological position supporting a project), of the issues and ideas to be explained, of the various factors that generate loyalty to the project sequence and to the project, of intellectual ways of presenting the ideology, of emotional ways of presenting it, of any great crisis that has been overcome in the project sequence history thus producing heroes, of simple ways of telling the project story, of the longer-term benefits of the project, and so on.

The ideology should advise people on how to approach the project sequence, protect the critical project stakeholders and, in some way, meet the psychological needs of participants. For instance, if the project is prestigious, emphasis might be placed on the quality of the work, on how the public will benefit and what a wonderful contribution stakeholders are making. In other words, project participants should feel that they are significant in making a unique contribution to their society.

The ideology needs to emphasise the important roles the different people play. It should be presented in such a way that it confirms the ongoing existence of the project and the willing contribution of participants in their roles.

Identifying the actors

People and organisations that take part in political and social processes are sometimes called actors. The word 'actor' has wider application than 'stakeholder'; every project involves actors. There are some very important actors in the environment, and these vary from project to project. A particular concern is to identify those actors who could have an impact on the project.

While it is obvious who some of the actors are, others may not be so obvious. People are very familiar with the role of local government in such areas as housing and office development, but may not be aware of other bodies such as courts and legal entities that operate on another level. (For example, in New South Wales, Australia, there is the Land and Environment Court.)

Actors derive their power from different sources: some derive it from the state and from the legal system. State power is probably more malleable than the power of the legal system. While there is enormous pressure to maintain the rule of law (or the appearance of it), there is much less pressure to maintain the rule of a particular government. Protest groups may have no legitimate power at all, they may in fact be illegal. The source of their power derives from coalitions of self-interest or coalitions sharing similar values. Protest groups usually need to maintain contact with their support base. Another source of power is expertise—professional organisations, while remaining within their area of expertise, influence people's views.

Where the actors are organisations, such as companies or political lobby groups, the group that actually makes the decisions is usually quite small (see chapter 3, the transition phase, pages 65–66).

Dealing with the external political environment

The work involved with the external political environment can be treated as a subproject with its own project manager and resources. The project manager has to carry out this public activity with limited funds, so priorities have to be set and judgements made about who is important and who is not.

The subproject needs to identify and focus on specific objectives. Clarification of the objectives is particularly important; this step comes before the decision on the methods to be used. However, the objectives may be modified according to the nature of the methods available. An objective need not be to win support but to neutralise the opposition to the extent they either give up or lose. A big win is nice, but a discreet achievement is often sufficient.

This subproject is essentially about the movement of information; there is no tangible product, except possibly a written document, tape or disk. The information is emotionally based—support and opposition are emotionally based—and the project's response needs to tap into emotions, though not necessarily in an explicit, blatant manner.

The project manager has to set up a process of data collection and decide how to use it. This area is so sensitive and changeable that good data is essential; once identified, it is wise to stay close to the controlling cliques.

On larger projects, specialists such as press officers, public relations consultants and advertising agencies are needed. Public meetings or appearances before legal and governmental inquiries may be involved. Communications with project supporters must be good in order to maintain and strengthen their support; this involves personal selling and may involve advertising. There will also be a great deal of contact with people both inside and outside the sponsoring organisation. It goes without saying there will be a lot of talking. Those authorised to deal with the media or other actors in the external environment need to maintain a consistent story.

How much to say

One dilemma that will emerge is deciding how much to say about the project. There are many who believe in telling the public as little as possible while others support full disclosure. Experience suggests that those in power prefer as little as possible to be divulged, and then only enough to deal with the particular situation at hand. This can pose a conflict between personal ethical standards and *realpolitik*. Either way, the project manager must try and assess how people would react to the various positions that could be taken and then choose between them.

If the position is taken where as little information as possible is divulged, the project manager needs to be certain that the opposition cannot make the accusation that data is being withheld. Credibility must be maintained. The public posture must be one of openness; secrecy leads to distrust (unless secrecy can be justified by claims that disclosure would prejudice the national interest, such as might be the case with defence projects). Another reason often advanced for secrecy is the need to maintain public order and to stop people panicking.

Sometimes, legal requirements govern disclosure. The law requires that various documents be produced and specifies the information they must give. It also defines what activities should take place, for instance, informing people who would be affected by the project, and so on. These documents and activities are important in the management of the relationship between the project and the various actors. They need to be carefully thought through in terms of their emotional impact. When dealing with these and other public processes, it is important to make the distinction between telling everything and telling the truth. But it is very important not to lie or, in machievellian terms, be caught lying.

In the political arena, there does not have to be a tight, logical connection between the action taken and the stated reason for the action—there only needs to be a plausible connection. For many project managers, brought up in the physical sciences, finding such tenuous connections don't come naturally; advice may therefore be needed. In practice, a small number of justifications cover a very wide range of actions—public safety, commercial confidentiality, fairness to all, etc.

Basically, what is needed is a good story; some might call it a good narrative. This is a very subtle art. To people locked into the physical world and seeing power and influence in physical terms, it might seem almost irrelevant. To keep the story going, the project manager needs a nice, neat, plausible one. This story, just as in the management of scope process, must give reasons why the project exists and how it is going to solve a range of problems. To justify the direction the project is taking, the story can be embellished with descriptions of the constraints on the project. It is also strengthened when emotive issues are brought in.

Conflict in the public arena

The main strategy in managing conflict in the public arena is to keep a good story upfront and defend it. The project manager does not have to win 100 per cent of public support; it is almost impossible anyway to get complete agreement in the public arena. Extraordinary things go on in people's minds, and the more one is involved in this area, the more one wonders that society agrees to anything at all. It is a great social achievement to gain agreement on which side of the road should be driven on!

There will always be disagreements in the public arena. Even within the protest organisations themselves there are not uniform views. It is incumbent on the project management to identify the range and strength of the views that exist, as this knowledge helps determine the project management response.

In liberal democratic societies, the right to protest is strongly entrenched. When one is dealing with international projects, the external environment becomes even more complicated, with clashing value systems, moralities and social practices.

The importance of project communication

Project communication is important and varied. As can be seen from the brief discussion in this chapter, project communication covers a wide range of skills and problems. For project success, good and effective communication is essential.

11

Chapter Eleven

Objectives and Scope Management

Keeping a project on track is very difficult. It is difficult to know what the project is about, to know what it is expected to be and to deal with the inevitable initial confusion. 'Okay, let's have an Olympic Stadium' is a very simple statement, but going from there to knowing what it is you want and how to get it is rather more complex—'Was that an 80,000-seat or 100,000-seat stadium?', 'Are there or are there not government funds involved?', etc. For the project to even take shape, many decisions have to be made. Then, once the overall form is determined, much work is needed to flush out the detail. In many cases, what the project is will not be clear until it is finished. In the making of a film, its precise nature is not known even when the script is prepared. When the script is drawn up, many alternatives will be open to the director. Probably not until the film is finished will it be known what it is like.

Agreed understandings

There must be, from a very early stage, 'agreed understandings' on how one will know what should in the project. Note the expression, 'how one will know'. Knowing the detail is not necessary, but knowing the criteria by which the project can be recognised is necessary. The agreed understandings need not be explicit, written down or in any way clearly visible, but they must exist; if they don't, the process of management of objectives and scope is impossible. The maintenance and development of agreed understandings requires considerable talking and interaction between people working on the project sequence.

The project story

Understandings are built up in people's minds based on what they think the project to be—they have a project history in mind, they know who is involved and have formed impressions of who and what they are, and they have experienced similar projects. In other words, each of the participants has their own 'project story' or their own 'project narrative'.[100] These stories or narratives are invested with meaning and feeling that will influence subsequent action. Agreed understandings are achieved when the participants' project stories or project narratives are compatible—not necessarily the same, just compatible. If the stories are incompatible, problems will be encountered on the project; incompatibilities will translate into confusion over scope and this will lead to inefficiencies, possibly fatal ones.

The interaction and the talking become less important as the project sequence proceeds, because more and more of the agreed understandings have been transferred to written documents, or have taken material shape in the project. The need for them continues, however, until the project is finished; in fact, agreed understandings take on a new life right at the end, at the commissioning and handover phase.

Managing knowing what is in the project requires processes which do not require all the answers to be known now, but to allow them to be found in due course. It also depends on processes where considerable knowledge about the project will never be written down but remain in people's heads. The following example will help to illustrate. In the film *Apollo 13*, a series of scenes deals with the declining level of oxygen in the spacecraft; there is a need to find a technical solution to the problem. Someone collects copies or duplicates of all the equipment and odds and sods that are on board and accessible. All this is thrown on a table; it is all the crew has to work with to find a way to make oxygen.

What has happened so far is enough for all team members to share a compatible story: there is agreement about the urgency of the situation, about the constraints and about the importance, and there is a high level of emotional commitment. From this point, it is possible for the viewer to understand what is in this small but vital project. The objective is clear; what is not clear is how to get there. What factors should be in the solution are easily recognisable, as are what factors should not be in the solution. A lot of work that was unspoken or unrecorded went on in people's heads.

The objectives coupled with the constraints are the criteria that lead to the definition of a boundary of what should be in the project. The agreed understandings are these criteria. As the project proceeds, the agreed understandings expand and clarify, allowing what is in the project to be correspondingly further defined.

Reaching agreement

When working on a project, it is surprising the extent to which many people agree as to what is in the project and what is not. At various levels of perception, they feel it, they hear it or they see it. While the process is not without disagreement, the emergence of this agreement is quite subtle and does not involve much paperwork. However, it does involve talking, and it involves telling and listening to stories.

Management of scope

One of the most important tasks the project manager has to do is to separate what is in the project from what is not—this leads to the concept of scope. Some factors contribute to the difficulty of this task. First, it is only as the project sequence proceeds that what is actually in the project is ascertained. Therefore, there needs to be a way of predicting what should be in the project. Secondly, as the project sequence proceeds, compromises have to be made; it becomes clear that what was wanted at the start cannot be delivered exactly as had been hoped. Gaps will emerge between what was originally wanted and the capacity of the project sequence to deliver. These gaps need to be managed. So there are problems of the unknown as well as problems associated with the practical being unable to meet the ideal.

Defining scope

'Scope' is an interesting word with a number of meanings. Though not used by all project managers, it is none the less a useful and important concept—or, more accurately, concepts.[101]

Everything encompassed by the project can be defined as its scope. However, it is not until the end of the project, or at least not until it is well under way, that what 'everything' is, is known. So, while a definition that says scope is everything that is in a project may be precise, it is fairly useless.

Another meaning is that scope is what any competent professional would expect to be in the project, even if it isn't spelt out in detail. Clearly, a building designer must provide foundations, and a photographer, film; these are things are clearly within their scope. A definition of scope where people anticipate the content of the project is of vital importance when dealing with some contractual disputes. These are discussed further under 'scope development' and 'change of scope' below.

Another possible definition is **'scope is a growing definition of what should be in the project'**. Scope is first of all a definition of what is in the project; this accords with the first definition above. But the problem is that the definition of the project develops as the project proceeds. This is the practice. Thus scope is defined as a growing definition. The next point is that the growing definition is forward-looking but takes on a judgemental tone. The 'growing definition' is of what *should* be in the project. It is essential to understand that scope is growing and developing. It is to be discovered as we work through the project. The 'growing definition' also requires the continuous use of judgement by the project mnager, particularly the client project manager.

But we need to go further with this definition. As the project proceeds, decisions, which are essentially choices, are made. With the idea of a 'growing definition', clearly these decisions become part of the scope. But how is the 'should' to be handled? This may seem a bit pedantic and why it is raised now is probably not clear. Suffice to say that the definition of scope needs to be expanded to:

scope is a growing definition of what should be in the project and will transform into a description of how the project is to be achieved.

The purpose of scope is to tell people what is in the project; it is a guide to where the project sequence is going. What this expanded definition accepts is that the definition of the project sequence may have to be part of the growing definition of what should be in the project.

An example could be where a building is required, of certain dimensions and characteristics. After initial work, a concrete structure is chosen as opposed to a steel building—a building was the purpose of the project sequence, not a concrete building. Having decided on a concrete building as the project, the definition of what should be in the project has been modified from 'a building' to 'a concrete building'. The decision on concrete will guide later work on the project, but it is not at its core.

There is hierarchy in scope—some parts of the scope are more important than others. 'How' the project is to be achieved is lower in the hierarchy than 'what' should be in it; the 'how' can be more easily changed than the 'what'. Nevertheless, 'how' becomes part of the scope. Scope is not objectives, it is what helps to meet objectives. Scope will be discovered as the project sequence progresses.

Developing scope

This is the elaboration of details required to achieve what is implied in the scope statement. How the requirements are stated is very significant. If they are stated in the form of exhaustive specifications (not only giving details of what is functionally required but details of how the requirements are to be met), little scope development can be expected. However, even with quite exhaustive specifications, some matters are forgotten, although any competent professional in the area would recognise that they had been overlooked, or their existence implied. If their existence is implied, these matters then give room for scope development. It could be argued that a specification requiring a building be designed without mentioning foundations still implied a requirement for foundation design (but maybe not—welcome to the world of contracts!).

A more open form of specification is the performance-type specification, which defines the function required and leaves the method of its delivery to the contractor. If the requirements are stated in performance terms, quite significant development can be expected. However, this does not completely remove the problem. An oil-change system can range from a single oil drum with a hand-operated pump to a fully automatic, technologically advanced extraction, purification and reuse system. A scope statement that simply said 'oil change system' is clearly inadequate. Although the budget allocation may imply some specific level of functionality, relying on that to indicate scope is not often satisfactory.

The implication is that if something is defined as scope development, then the supplier bears the cost of that development. The client project manager may wish that scope development covers everything while the supplier project manager may argue that a 'change of scope' has occurred.

Changing scope

A change of scope occurs when what is now expected is materially different from what was previously reasonably expected. Changes of scope can arise because fundamental data received from the client has had to be altered. Examples are increasing the functionality of a piece of software, or the introduction of an overhead travelling crane into a factory.

The big danger with change of scope is that the visible, obvious change may only be a small part of the overall change. The logical follow-ons from the change also need to be considered. In the example of the overhead travelling crane, its introduction increases the height of the building, increases the downward load on the foundations, increases the area of building exposed to wind loading, increases the cladding requirements, and so on.

Ratios of indirect to direct effects of changes of scope in the order of five to one are not uncommon, but are rarely fully recognised.

It is important that, while the client project manager should strenuously resist changes of scope, the need to cope with them be addressed. The need to change scope can arise from some human failure to identify what was required earlier. It is also a normal consequence of the nature of projects, where information about the project increases as time goes on; the least information is available at the beginning of a project.

It is often difficult to distinguish between scope development and change of scope. For instance, are the software manuals written in plain enough language? And if so, how has this been determined? The client project manager should be wary, because it is easy to accept scope development under the guise of change of scope if there are plenty of funds to meet the change. Making changes early on in the project may use up contingency funds needed later.

The client project manager and the supplier project manager are on different sides of the fence here. In some situations, enormous pressure is applied to the supplier project manager to find changes of scope. This happens especially when the contractor has a low or narrow margin on the job; the contractor bid below cost in the expectation that the difference could be made up later when the project had started, by means of changes of scope.

The supplier project manager starts to put great effort into arguing that any change is a change of scope, and not scope development. It is this kind of behaviour that leads some clients to be concerned, right from the outset, that the supplier had allowed itself an adequate margin so that it can concentrate on doing the job and not spend time finding loopholes. *Partnering* is an attempt to set up a relationship to minimise loopholes. It requires serious effort on all sides, particularly client representatives, to be clear on what is required and expected. It is the need to be both willing and able to put in this effort in the early stages, when least is known, that makes the client's role so onerous.

In some industries, a change of scope is called a variation.

Accepting the scope statement

Keeping the project on track is very difficult: it is difficult to know what the project is about and what it is expected to be, and, following from that, it is difficult to deal with the inevitable confusion. The progressive elaboration of the project moves from one phase to the next phase.

As often happens on projects, a new team is assembled to continue with the project; this is called 'over the wall' project management. It is as if the people who worked on the feasibility phase took all they had developed and tossed it over the wall to the next team without much handover work. That this happens is regrettable, but it does and there is a need to know how to deal with it.

The opening sequence for the management of scope is to achieve an agreed understanding of where the present definition of the project scope stands. Much of this can be achieved by a review of the document or documents that contain the functional specification of the project.

Controlling changes in scope

There is always change on a project—no sooner have people agreed on what the project will be, have agreed what will be in it, and have agreed on how it will work, they start making changes. It is quite acceptable for scope to change, but only as long as it is agreed to by the client. Change is a healthy part of project life, but it needs to be controlled.

An added difficulty on long projects is creeping or gradual change, which can be difficult to identify. The project manager should therefore, at regular though reasonably large intervals, set some markers or statements of objectives that can be referred to in order to identify any large, but gradual, changes.

As changes of scope come from both internal and external sources, a formal process of management is required. This process is usually one of variation management or configuration management.

Variation management

The project manager must set up a procedure by which the changes of scope and scope development are managed. The system has to have a formal mechanism by which a change is proposed, by which it is recognised, by which it is approved or rejected, and by which the decision is communicated to the interested parties. The imposition and maintenance of this management process is the only effective way to protect the project from the possibility of changing priorities and objectives. Essentially, if anyone wants a variation and can pay for it, they can argue for the change; otherwise the variation is not entertained.

Proposals for variations may come from any source, and the client project manager must put in place a procedure for their central collection. Without central collection, the element of control is lost. Typically, that central collection point numbers each proposed variation, which is registered in a document or a computer file.

Any directive from a client that a variation be introduced is approved and allocated a number. Requests from suppliers for variations are collected and numbered. Supplier requests may be as a result of conflicts in directions given to the supplier; they may be suggestions for improvements; they may arise simply as claims for more money, based on the assertion that what the supplier is now being asked to do is not what was agreed originally.

It is important that, whatever the source of the request for variation, it be logged.

The client project manager makes a recommendation to the client who decides on the legitimacy of the supplier's request. If it is rejected, the supplier may appeal. The appeal against the decision may be raised in meetings with the client or taken through various mediation routes, including the courts. It has happened quite frequently that the legal costs have vastly exceeded the original cost of the variation.

Since there must be some way people on the project can find out when a variation has been approved, a register of the status of variations must be set up. A decision to approve a variation and accept it as a change of scope usually carries with it an increase in budget and time.

When a variation has been submitted, there is general pressure to put the whole area affected on hold. This is not recommended—work should proceed under the approved scope. What has been approved and accepted should not be compromised by what is only a possibility of change.

Configuration management

The most highly developed variation management procedure is configuration management. The aim of this is to keep the project on track and know where it is at any point in time; it also keeps track of documentation and specifications. Configuration management has concepts expressed by such words as baselines, baseline management, functional baseline, allocated baselines, design baseline, product baseline, operational baseline, traceability, configuration control board, configuration auditing, and configuration status accounting. On any project of some complexity in a changing environment, configuration management is essential and needs to be set up in its formal form.[102]

For simple projects, full-blown configuration management would be too expensive, and regular design reviews are enough to keep track of scope. Such reviews, however, should have a formal system for numbering, listing and approving or rejecting variations. The control mechanism does not have to be complex but it should be visible and clear. Both client and supplier should be party to design reviews.

For a project of any reasonable size, configuration management has a role. At present, its more developed use is limited to a small number of industries, particularly defence, but there is opportunity for much wider application.

The impact of proposed changes must be examined in configuration management. Impact is not always obvious, so it is advisable to involve the designers themselves in trying to identify the impact of proposed changes. The changes will affect the work of the designer and the designer's organisation, and, more than likely, the work of outside designers. Not only will individual designers themselves have to be satisfied about the proposed changes, but also other designers and interested parties.

Defining objectives

To keep things on track, there need to be some signposts. These signposts always lead eventually to people, to the client and to the stakeholders. Initially, the signposts are the objectives to be met by the completion of the project. There can be much difficulty and conflict in getting these objectives in place; getting them in place is one of the functions of the transition phase. The problem with objectives is that, as work on the project moves along, there is not a complete one-to-one link connecting the objectives developed at the beginning to the objectives as they develop through the life of the project. Two things happen: one, the objectives at the beginning are broader than the specific solution that emerges at the end of the project sequence; and two, gaps between what was wanted and what is possible emerge. It is in the management of this broad-to-specific aspect of project development and in the way the gaps are dealt with that the real objectives and scope management activity are concentrated.

When objectives are stated in the beginning, there is always a wide range of project deliverables that would satisfy them. A declining market share may prompt a company to take some action; it may have the objective of recovering cash flow. This objective can be met by a range of strategies, each strategy representing a project outcome. At the generic level, there is product improvement, change in price, change in product, change in the market, and so on. The greatest range of possibilities occurs at the beginning; however, as the project takes form, the possibilities become fewer. At each point in the process of narrowing down the options, one must attempt to satisfy the objectives. In a very strict version of project management, the narrowed path must always satisfy the objectives as originally defined; this may in fact be impossible because of some unreasonable expectations. Added to this is the problem that the objectives themselves may have to change, as more data on their viability becomes available. Note the possibility of changing objectives!

Changing the objectives needs clear endorsement from the client and key stakeholders. There is a danger that an uncertain client will bend too easily to pressures for change, or that a pigheaded client will refuse to see the writing on the wall. Irrespective of what it is—weak, pigheaded or sensible client—because everything is in the future, there is no assurance of doing the right thing when changing objectives. Project management culture is very strongly tied to the 'no change' position, a general position this author recommends. Pressure to change objectives should receive very close scrutiny.

Selection of objectives

This is not always as simple as it seems. There often is a sequence whereby one goes from gross and bland objectives to scope, to a new definition of the objectives and then to scope and so on. Gross but quite bland objectives are fairly easily developed; however, translating these into objectives that can give direction to an appropriate project requires effort and skill.

An example might be a company that wants to lift its sales; it has a feeling that something needs to be done, but what? There are in fact a number of generic strategies to identify this. A vague feeling that 'something needs to be done' leads to a subproject aimed at finding an appropriate set of objectives through which to get the company out of trouble, and a determination of the scope of the subproject. When these objectives are recognised (and they might be to solve a quality problem), progress can be made towards a new description of the scope. In this case, the scope will be all the work necessary to solve the quality problem, which might lead to ISO quality standards accreditation.

Objectives and scope document

Most projects have a key document describing the objectives and scope of the project. it may be called the project manual, the product requirements document, and so on. This document give justifications for the project along with a record of the definition of the work to be achieved. In practice, only one version of this document is prepared and becomes a major reference. It also describes procedures to be adopted for the approval of changes and nominates levels of authority.

In more complicated and larger projects, this document needs to be seen as a growing and developing document, the formal reference of where the project stands. Earlier versions should be held to allow reviews to take place.

Project control framework applied to scope management

It is useful to illustrate the process of managing objectives and scope through three phases. The phases selected here are transition, feasibility and planning and design. Note that the objectives plus the elaboration of the scope in one phase is passed on to the next phase to be elaborated by a further definition of scope and, perhaps, modification of the objectives.

Managing objectives and scope in the transition phase

This is the phase when the project is conceived. Out of it emerge the objectives for the project. There is no obvious source for the objectives, except that they spring from personal desires. As far as this phase is concerned, there is no control framework: it will produce a set of objectives which will be passed on to the feasibility phase.

In the movie world, the acceptance of a script (or even a story-line) by a group of investors might be the event and outcome of the transition phase. The objectives are to make money and make a movie from the script. A constraint is the amount of money available. The movie industry is probably unusual in that making a movie can be its own reward. In any event, the objectives have been set in a broad framework.

These objectives are passed on to the feasibility phase. (Note that this care and attention to the phases may not happen in practice, but the problems faced will have to be solved either formally or informally.)

Managing objectives and scope in the feasibility phase

Taking the movie example further, we now have objectives (say artistic merit is present), a script and money. The script may have to be changed but what we want to do is establish that we can make this movie for the amount available and, further, that there is a good chance of making a profit, if this is one of the objectives. These are not easy questions to answer. A balance must be worked out between the requirements of the script and the resources available to meet them.

In managing scope, one needs to keep all the activity and develop the work within the context that there is a script and a certain amount of money. A new script is not suddenly taken up, or the assumption made that the budget has doubled. The requirements of the script and the technology of production lead to a number of decisions, which now constitute the scope of the project.

There now comes scope as the definition of what should be in the project (the various scenes the movie will have), plus scope as the definition of the means of achieving what should be in the project (which filming technology to use). There is also the fine line between objectives and scope. The method of production may influence the artistic values achievable.

During the feasibility phase, the objectives have been converted to a statement of scope, and staying within this scope becomes the objective for the next phase.

Managing objectives and scope in the planning and design phase

The feasibility phase has converted objectives to a scope statement or, to be more precise, the feasibility stage has taken an objective and added a scope statement to it. The scope statement is a description of what is to be in the project. It may also contain a description of how that is to be achieved. The feasibility phase answers or provides some answers—it leads to the first formal definition of scope.

In the movie, the script is detailed further, the scenes are further developed, locations are considered, and much more detail on exactly what will be in the movie and how it will get there is developed. These details and decisions are passed on to the next phase as the controlling or guiding directions.

Development of objectives

The general sequence in the development of objectives that will ultimately lead to the realisation of the project—and beyond—goes something like this:

- wants and needs (for example, economic growth)
- lead to the statement of an objective (more electrical power is needed)
- which leads to a specific proposal (nuclear electric plant)
- which leads to an actual project (a nuclear plant)
- which leads to something that satisfies the objective (electricity as a source of power)
- which leads to, or possibly might lead to, economic growth.

However, beyond all this, the plant leads to considerable political disruption and the emergence of other political forces; it is also blamed for birth defects.

The above sequence can be varied and presented in different forms. Basically, there is a need for something; steps are taken to try and get it, which may or may not succeed; and in all probability there will be unintended consequences to deal with.

Reaching agreement on objectives

The objectives at the very beginning are by necessity broad; they are then reduced in range. What is important to note is that the broad wants and needs do not lead logically and necessarily to a specific proposal—a corporate objective to make money does not lead to the inevitable conclusion that a piggery should be built. This fact makes the project objectives issue harder to manage. What should be done is to try and gain agreement at the end of each of the steps. The project manager then needs to hold the participants to the understandings behind that agreement. Note the word 'understandings'; it is these understandings, many of which are unspoken, that are the clue to holding the project on course.

Arguing the technology

At the outset, it is appropriate to sound a word of warning to project managers—stick to the management. The project manager is appointed to manage, and that means getting work done through others. It is necessary to reiterate the warning because, for many project managers, they are on home ground with technology and it is all too easy to gravitate towards their area of basic qualification. It is important that project managers resist the attractions of technology—success in project management is more likely when the project manager stays with the role of manager. However, this does not mean forgetting or ignoring technology. The challenge for the project manager is how to approach the technology and how to manage it without getting immersed in it.

Without getting bogged down in a deep discussion on its place and meaning, technology is, essentially, the use of specialised data or information. There are other aspects, such as machines, chemicals and computers. Their existence is a manifestation of technology, not an exclusive definition of it. For example, when a surgeon uses a scalpel, the high technology is unlikely to reside in the scalpel.

This broader definition of technology allows us to recognise that projects that are essentially of a social nature do have a technological element. An example might be an organisational change project. Here, techniques based on certain theories and practices are applied in an attempt to achieve certain ends. Factors other than technology are involved in organisational change, things that might bring the whole project into question, but there still is a technology. The effect for the project manager is the same—while there is a technology which the project manager may or may not be skilled in, or even understand, it still has to be managed.

Assessing risk

Many projects experience problems with technology, particularly new technology. Well-established, proven technology is rarely seen as a source of serious threat to the project, whereas unproven technology is most definitely seen as a threat. A project with a large component of unproven technology is extremely vulnerable. One way to reduce the risk might be to break the project up into a number of subprojects where each contains a low level of new, unproven technology. For example, the huge NASA project to get to the moon was broken down into a series of smaller projects.

The emergence of specialists has had a significant effect in lowering the general level of risk associated with a particular technology. Specialists are able to deal confidently with their own technology, so specialisation, which arises for a number of other reasons, also helps in risk reduction.

When the technology of the project is outside the knowledge of the project managers (which will be partly true on most projects), they need to establish procedures by which to identify possible problems. This is all very well when the relevant technical expertise is available and can identify areas of concern. Unfortunately, sometimes on projects there are conflicts of interest, concerns with liability, or even a gung-ho approach, all of which can lead to expressions of concern about the technology being suppressed. There are a number of ways in which the problem of identifying risk associated with unfamiliar technology can be approached.

The project manager should assume that the technology does not work and demand proof that it does work, ie, proof that the technology works. This is quite a different position from assuming that it does work and then asking for proof that it does not—an inherently dangerous position.

The project manager may also reserve the right to have access to the internal documents produced by designers and contractors and send a project team member to their offices on a regular basis to read them. It is likely that any concerns the designers or contractors have will surface in their memos, minutes or other documents.

The project manager can also decide to get a second opinion. A fourth option is to seek out and identify where changes occur; it is at these points that problems are likely to be encountered.

Dealing with technology outside one's skill-base

There is often debate over the question of whether or not people can manage technologies outside their skill-base. The answer is clearly yes, as many managers do it day after day in the normal running of business, and they do it because there are one or two layers of management who translate the technical data into more common or generic language. These intermediate layers translate the technical language, with perhaps some loss in the translation, into a form that upper management can understand. As long as these layers are available, managing outside one's skill-base should not be difficult. This is, however, a fairly expensive solution.

At every level in the management of most endeavours, there are people who know and understand the technology of the business. While collectively they know it all, each individual knows only a part of it. The management structure, through layers of management, links these people so that a comprehensive view is possible.

It is somewhat similar on projects, but with a key difference. On every project, there is a group of people who collectively know and understand the technology, with each only knowing a part of it (there can also be parts of the technology about which no-one knows). On most projects, however, there is not enough room for a few layers of management to translate the technology for the benefit of upper management. The project manager has to directly deal with the specialists. It is because of this need to deal directly with the specialists that people believe that project managers need a strong background in the technology of the industry, or at least in the dominant technology. (Although the author regards this as an open question, he does want the surgeon to be highly technically qualified and not simply up there managing!) Even if the project manager has a background in one of the technologies, there will be others that are unfamiliar. To deal with this problem, the project manager needs to:

- learn the minimal amount of technical jargon or language;
- identify the absolute core of the technology and develop an understanding of it;
- identify the key control technologies associated with the core technology;
- know what documentation is used by the specialists and how it is used; and
- identify reliable experts to which one can relate and get information.

By combining the above with the following four-question module (and by continually asking other questions), one can make a good fist of managing technology outside one's skill-base.

- What did we decide to do ?
- What was done?
- What is the difference ?
- Why?

Dealing with jargon

Every technology has its jargon, much of it being simply other words for the same things found in other industries and specialties. A small number of terms in this vocabulary contain the central concepts. Minutes of meetings and dialogue between experts are littered with jargon. The words can also be encountered on a visit to the library or during discussions with colleagues who are not involved in the project but have expertise in one of the technologies being used in it.

The experts have absolutely no expectation that the project manager knows anything about what is going on at the technical level, and they can be left in this state of misapprehension to be surprised now and again by a question. It is pointless for a project manager to get into arguing the technology—the experts know more. However, it is highly appropriate for the project manager to find out what impact the technical arguments have on time and interfaces. The experts quite happily acknowledge the project manager's concerns in areas that come under the project manager's jurisdiction, but resist project management interference in their technology.

The big advantage of talking to anyone skilled in a certain technology is that they are able to prioritise the data and indicate the main areas for concern. In most technical disciplines, there is only a small number of areas of concern; the rest is usually covered adequately by normal practice and people feel quite secure about it.

Controlling core technology

Core technology always has control systems associated with it. It is useful to separate the technology from the control systems.

Core technology is the key technical process that achieves the purpose of the project. If it is an organisational change, it will be seen as an overview strategy—unfreeze, change, refreeze, for instance.[103] In energy generation, it will be the technology of the source of power—a coal furnace and steam, a hydro-electric motor and a dam, or a nuclear reaction with rods and cooling. It would be possible to summarise the core technology in under 10 pages; identifying the science behind it and a statement of significant points on its operation is sufficient.

The core technology is a very small proportion of the project, less than five per cent and possibly less than one per cent of the total technical content. For example, the core technology of the automobile is the engine. The rest of the engine is controls. The most significant parts of most cars are the shape, the upholstery and a few gadgets! The project manager can confidently assume that the principles of the core technology are limited in scope and can be mastered in a few weeks on the job with a little judicious reading and talking.

There are also control systems. The core technology rarely works easily and many control features have to be built in for it to work. Often the full extent of the control systems required is only discovered after repeated failures and frustrating experiences. This is the big danger with new technology—underestimating the extent of the control systems required.

The technology underpinning the control systems relies on a limited range of basic technologies. A steady reading of the principles of these will give the project manager a store of technologies to be taken from project to project. Remember, project management involves problem-solving over time; the project manager has a certain amount of time in hand in which to come to grips with the basics of the technologies.

There is often more technical content in the control systems than in the core technology. The viability of the internal combustion engine probably lies in the lubricating, the cooling and the electrical systems—the systems that control the operation of the core technology. Without them, the core technology simply overheats, melts or doesn't work at all. In managing technology, therefore, the project manager can expect to spend more time on the control technology than on the core technology. In the Three Mile Island disaster, it was the control technology that failed, not the core technology.

The technologies underpinning the control systems emerge strongly during the planning and design phase of the project and again right at the end at the commissioning and handover phase.

Managing through experts

The technology is managed through the experts. The project manager must find them, appoint them and come to an agreement with them on what is expected of them. Experts should provide the project manager with regular reports on matters within their jurisdiction, and supply recommendations if so required. Their recommendations become the basis of action by the project manager.

Staying focused

In order to remain calm and in command of the situation, project managers need to develop some way of staying focused on the project sequence; they need to psychologically protect themselves from the confusion and imprecision that goes with the territory. This protection comes with the understanding that a sequence of progressive elaboration is involved in all projects. It is important to recognise that as the project sequence proceeds, more and more data on the project will come to light, be brought forward and turned into information. In fact, it is vital to recognise that the description of the project and the sequence of developing more and more detail must take place over a period of time, and that, in many ways, there is a natural or sensible order in which the details develop. Starting to shoot a film before some vital plot details are sorted out is a sure way to waste money.

Project managers should therefore develop a personal reminder program of times when certain details of the project will be available or should be available. This program will be related to the project life cycle and will reflect the fact that the elaboration of the project is spread out over time. Such a program allows project managers to focus effectively at any point during the project cycle on the things that are important just then.

A key activity

Managing scope, objectives and their associated technologies is a core activity of project management. Success or failure of project management can often be traced back to success or failure in managing objectives and scope.

Chapter Twelve

12

Interfaces

Despite its importance in project management, interface management has been given little attention. This might be because it is difficult to describe it succinctly without appearing trivial. Also, it is so pervasive in project management that no-one really notices it, rather like the air we breathe. But people are beginning to give it more attention.[104]

Interface management must be done, and done by the project manager. No other project participant has responsibility for it; the demands of interface management mitigate against any one of the specialists doing the job. If there are few requirements for interface management or a project is completely dominated by one technology, consideration might be given to handing the interface management over to a specialist. In most projects of any reasonable size, however, the requirements of interface management are such that a project manager is needed.

Interface management consumes much energy and skill. It is a sea of detail that needs to be managed and simplified wherever possible. Much of the detail is often beyond the immediate understanding of the project manager.

The project manager chooses and imposes interfaces. Time interfaces may be imposed to ensure a certain kind of work is finished before another starts; geographic interfaces may involve taking work off-site so as to allow other production processes to commence; technical interfaces may be imposed to ensure that certain technologies come together; and social interfaces may be imposed to keep certain work-groups apart. In contrast to these interfaces, where elements are separated, elements may be brought together in an attempt to reduce the number of interfaces. The most striking example of this is where people are brought into a project all in the one place.

Good interface management is crucial if the project is to succeed, otherwise there may be project failure. It has been claimed that a satellite launch blew up because a cleaning cloth left by one subcontractor was not removed by the next subcontractor.

Overview of interface management

It is not that easy to get a watertight definition of an interface; also, the definition needs to change to suit different contexts. However, the following definition sets out in the main the conditions that define an interface of interests in project management.

An interface is a boundary where an interdependency exists across that boundary and where responsibility for the interdependency changes across that boundary.

A boundary itself is a necessary but not sufficient condition; an interdependency itself is necessary but not sufficient; and the combination of a boundary and an interdependency is also a necessary but not sufficient condition to define an interface. To achieve an interface of interests in project management, there has to be the added condition of divided responsibility. (For the logically minded reader, it is admitted that an interdependence implies the existence of a boundary, but a boundary does not imply an interdependence. Even though this is so, there is a need to recognise both the boundary and the interdependency in the definition.)

In practice, the conditions that define the boundary are usually quite different from those that define the interdependency. Usually the boundary is determined by the arrangement of people or organisations; the conditions determining the interdependency are usually based on technology. For instance, in wiring one electrical item to another, there is interdependency and a boundary. If one person is doing the wiring, we do not have an interface in terms that interest project management (there is an interface in other intellectual frameworks but not in the project management framework). For this to become an interface in project management terms, there must be one person responsible for the wiring to one piece of equipment and another person responsible for the wiring to another piece of equipment.

Interface management then is the management of the interdependencies and responsibilities across the boundary of the interface. The need for interface management grows as that interdependency increases; its focus is on the behaviour of the participants and the behaviour of the technology at the interface. Interface management is a function that has to be resourced throughout the project sequence.

Sources of interfaces

Interfaces arise because work is broken down into parts and each of the parts is carried out or executed by different people or organisations. The definition of the parts may depend on technology, on economics, on geography or a host of other factors. One source of interface is social, that is, an interface is set up because of social forces or social reasons rather than any strong technical basis. The project manager can also impose interfaces to help in the management of the project.

The greater the interdependence, the greater the need for interface management. A whole range of matters can be managed across interfaces, such as operational requirements, specifications, drawing conventions, measurement systems, colour co-ordination, language conventions, continuity, word processing systems, CAD systems, data structures, report formats, decision processes, allocations of authority, co-ordination of space, infrastructure issues, political policies, disciplinary procedures, laws governing contracts and supply times.

No matter how the work is divided up, there is always the problem of linking the various parts and so a need for interface management. If the work is all being done within one organisation, that organisation must link the parts; this is internal interface management. Where work between organisations is to be linked, external interface management is needed.

Interfacing the client

Now we turn to the work of client project managers in their role of managing the client, making sure that the client does what is required of it and continues to support the project.

Supporting client decision-making

The client project manager must first of all keep the client generally well informed. This is done by providing the client with a regular project report on events and on such key project benchmarks as time and cost. The report should highlight decisions to be made in the coming period. To the busy project manager, this may seem unwarranted effort; however, it is essential. In practice, the first report takes the longest to prepare; from then on, the same format is usually followed, with issues carrying over from one period to the next. The reports, when reviewed together, highlight the significant issues affecting the project. Each report gives the project manager a feeling of progress and is a useful discipline. More importantly, however, it assists the client.

One of the real frustrations for participants in a project is when client decision-making slows to a point where the delay in making decisions is holding up subsequent decision-making. Such decisions might concern a sequence of variations. It becomes very difficult, when one is trying to consider the impact of a proposed variation, not knowing which outstanding variation proposals will be approved.

The client project manager needs to identify the decisions to be made and to collect the appropriate data. If analysis of this data is possible, it should be provided. The purpose is to make the client decision-making as easy as possible and to reduce delays. Morale also goes up when the client is seen to be on top of the job.

In presenting advice, the project manager will have to rely on, and closely interact with, experts. A standard way of supporting client decision-making is for the project manager to have a report prepared with a set of recommendations. The client is then asked to adopt the recommendations. Another way is to outline the alternatives so that the client can choose which direction to take, that is, choose between the alternatives. This all looks simple enough, but it really is only be the tip of the iceberg. Clients need to be prepared for these decisions—they should not be sprung on them (unless one is into a very interesting and dangerous political game !). People need time to make important decisions, they need to weigh up the pros and cons in their own way. Sometimes they need notice that a particular problem is coming up, long before it formally appears in a report with recommendations, so that they can prepare themselves.

On the following project, known to the author, a little more effort should have been made to keep the client informed. The number and frequency of subcontracts to be signed gave the client the impression that the project was out of control. The contracts were to be expected, were appropriate and were in budget, but they just kept coming and coming, committing the client to more and more expenditure. Although the project manager knew that they were expected, and knew that they had budget allocations, the client did not. Supporting client decision-making is a vital role of the project manager.

Ensuring client support for the project manager

Once obtained, the project manager must work to keep the client's support. This means putting thought and effort into client relations and ascertaining the client's perception of the role of the project manager. This is a classic managing up situation: the project manager must find out what is important to the client, ensure that the client's need for certain information is satisfied (even if it appears irrelevant), advise the client of what the project manager intends to do and gain agreement from the client, and keep surprises to the minimum. The old adage of not bringing problems unless one also brings a solution applies.

The project manager must try to win the trust of the client. In this context, trust means that the client can rely on the project manager to conduct things in a manner the client deems appropriate. Regular, reliable and frequent reporting and recommendations are a key to building that trust. Project managers should see themselves as members of the client's team, acting and advising the client. There is an interesting tension here between the needs of the project and the needs of the person or entity in the role of the client. There are of course personal dynamics involved in the relationship between the client and the project manager. The nature of these dynamics need to be evaluated—they may be positive or negative.

Dealing with a complex client

Where the client is a complex entity rather than just one person, project decisions are shared decisions, and the project manager has to manage the support and decision-making of an interacting group of people. A project control group should be established to facilitate project decision-making, otherwise it can become very difficult and frustrating trying to manage the client interface.

Managing internal interfaces

Internal interface management in an organisation with a functional hierarchy might be carried out by vesting the appropriate level of authority in a co-ordinating manager. The co-ordinating manager has the authority to act and to direct people. Alternatively, a project manager responsible for the whole task or subproject might be appointed. However, in this case, authority is rarely clearly and unambiguously in the hands of the project manager.

The project manager approach to internal interface management can be illustrated by the following. In the position of product manager, that manager is responsible for looking after the work necessary to support the product's manufacture and distribution. Such a person would not have the ability to direct work within departments but would be concerned with the work's progress through departments and how it is passed on to the next department. The product manager would be responsible for all the interfacing work and ensuring that the product got the correct support from the departments.

Internal interface management requires considerable negotiating skills. These skills must be practised within a context of long-term, ongoing relationships within existing hierarchies. The project manager needs to gain the resources and have control over their direction while still maintaining relationships that will be needed on the next project and beyond.

Some organisations attempt to put the actual power into the hands of the internal project managers and shift it away from the departmental heads. This is done by an accounting device whereby the project manager is seen as having the money; this money is the internal project budget or the revenue expected from the products of the internal project. In addition to being able to buy the services of the departments, the project manager supposedly has the power to go outside the organisation to buy the services and bypass the internal departments. This process seems simplistic to the author and can be quite dangerous for an organisation, but it does strengthen the hand of the project manager.

Maintaining external interfaces

There are many sets of external interfaces. For the contractor, there is the management of the interfaces between the subcontractors and between the subcontractors and the contractor. For the client, there is either simply the relationship with the turnkey supplier or the interfacing between the contractors. Just as we have many project managers, we have many people managing external interfaces.

The magnitude of this issue of external interface management escalates geometrically rather than arithmetically as the number of organisations increases. Between two organisations, there is one interface; between three, there are three; between four, there are six; between five, there are ten; between six, there are 15, and so on. Once the numbers become reasonably large, the number of interfaces is potentially impossible to manage. Mega-projects have something in the order of 200 subcontractors. In practice, many possible interfaces don't exist, simply because of technical and time separation between subcontractors. In any event, the project manager may have to introduce artificial boundaries with special management provisions to manage through these boundaries. The introduction of project life phases is one such device to reduce the number of open interfaces.

The general principle in project management is that any contractor, subcontractor, consultant or other supplier is responsible for what goes on inside their own organisation, and that project managers are responsible only for work that falls between the organisations. Although this state of affairs is legally desirable, it is not sufficient to ensure project success; it may be sufficient to allocate blame in the case of failure, but it does not help ensure success. A project manager charged with achieving a successful project would almost certainly need to be assured that adequate internal interface management was in place.

Whereas with internal interface management, there is some scope for organisations to increase the power and authority of project managers, there is less opportunity to increase the authority of the project manager charged with managing external interfaces. High-level negotiating skills are required, as well as other persuasive personality characteristics.

One might think that a contract is sufficient; while the contract should certainly assist in promoting the authority of the project manager, it in no way guarantees it.

An attempt to overcome the weakness of contracts in the area of interface management and develop more socially based techniques is embodied in what is called 'partnering'. Partnering has many meanings. In its most complete form, it means that a group of organisations work together over many projects, and that this commitment to a longer-term relationship is used as a basis for interface management. There is the possibility of developing compatible specifications, and of using a particular language and particular drawing conventions. Such standardisation across a number of organisations would reduce interface problems, and other advantages would be derived beyond interface management.

One variation of partnering is a commitment from people to work together in a spirit of co-operation for the benefit of the project. This idea has some legal implications embedded in it that have not yet been clarified. While this model of partnering has had its successes, it has probably also had its failures.

Types of interface

In a generic sense, there are three interface conditions: perfect match, partial match and total mismatch. In practice, there are likely to be combinations of these conditions.

Managing a perfect match

In a perfect match, there is perfect matching across the parts and everything meshes exactly. It is the objective of interface management to achieve this situation, where the interdependencies have been worked out so that everything exactly fits together. Components are physically and operationally compatible, data transmission is seamless, and so on.

Managing the partial match

A partial match happens quite frequently and can often be masked by an apparent perfect match. This is actually a very difficult situation which can be quite dangerous. There is some commonality of work practices, specifications or maybe even values. It appears as a lesser problem but it can still cause serious difficulties. The usual problem for project managers is to identify the mismatch and the degree of mismatch.

Managing the total mismatch

Here work practices are different and there is no agreement as to how things come together. It is the usual initial situation found by the project manager, who needs to move it to the perfect match situation. The problem can become one of jurisdiction over the interdependency; for example, where two units or entities each think they are in charge of the same thing, or they each set the standard and the other complies.

Managing interdependencies

In practice, at any one boundary there are likely to be many interdependencies. Some of them may meet the perfect match situation while others may be a total mismatch. Some interdependencies may be a combination of all three: for example, in relation to drawings, there may be no agreement as to the meaning of symbols on drawings (total mismatch), but total agreement on measuring conventions and measurement systems (perfect match). So at each boundary, different conditions may apply to different interdependencies.

Interfaces are always in some way related to people, organisations, professional/skill disciplines or time. The dominant determinant of the boundary is discipline rather than people and because of this, considerable interface issues can be thought through simply on a discipline basis thus allowing some preplanning before people join the project. On the basis of knowing what disciplines will be involved, the project manager can identify some of the interface issues to be managed, and maybe avoided.

Interdependency across an interface may function well, but it may also involve interpersonal conflict. The project manager then has to play a mediating and conflict-resolution role, which can be quite difficult.

Managing time interfaces

There are interfaces over time, between one part of the project sequence and the next—there are logical and technical reasons why these exist. There are also interfaces which are introduced by management. Interfaces in time are given a great deal of thought in quality management; the concept of the downstream customer is introduced so as to focus people on what they should actually pass on and how they should do it. Time interfaces are affected by the degree of continuity in the involvement of the participants.

Management of the interfaces

As can be appreciated, there is a lot of work in getting all this together. While any one of the interface problems may seem easy to deal with (and most will, though some will be intrinsically complex), the big problem facing the project manager is getting them all right (or at least the important ones right) at the right time. On any sizeable project, any solution to interface management must recognise the need to get as many details as possible right at the same time.

A difference between process and project management might be highlighted here. In process management, the interfaces also need to be identified and managed. However, as they are of a more permanent nature, there is opportunity to refine their management. In project management, the second chance does not exist to anything like the same extent and the interfaces have to be managed using transient and temporary organisations.

Identifying interface boundaries

The first step in interface management is to identify the boundaries and then the interdependencies. The Work Breakdown Structure (WBS), Product Breakdown Structure (PBS) and Organisational Breakdown Structure (OBS) are useful devices in this process. They identify some of the participants and some of the interfaces. These structures can be used in association with the interface matrix method described below.

Projects require a detailed procurement plan. Amongst other things, this plan defines the procurement strategy and identifies what service or thing is to be obtained from what supplier. Of interest here is the guidance it gives to identifying interfaces—by specifying the suppliers and what is to be supplied, the plan helps identify potential interfaces.

Drawing up interface matrices

The identification of interfaces between organisations, people, disciplines and time may be more easily identified using an interface matrix. The matrix has the same list of organisations, people and disciplines down the side and across the top. Their degree of interdependence and relative importance can easily be identified by coding at the intersecting points on the matrix. The matrix should be symmetrical with no interfaces along the diagonals. One might be a bit more sophisticated and use one triangle of the matrix to map the degree of interdependencies and the other triangle to map the information that is to be managed. Resourcing requirements can be quickly ascertained using the matrix as a starting point. The interfaces might be separated out on a time basis. To do this, an interface matrix can be drawn up as it applies, at the minimum, to each of the project phases and at the handover from one phase to the next. It would also be useful to prepare the matrix for the period of critical activities; in this way, the importance of the interfaces over time is highlighted. By doing this, the project manager can spread the resources over the interfaces over a time period, rather than all at once.

Identifying the nature of the interfaces

The description above on the types of interfaces is relevant. The critical interfaces, the ones likely to cause most damage if they fail, should be identified—the interface matrix is useful here.

Putting resources into serving the needs of the interface

Having identified the interface and the relative importance of the interface, decisions on resourcing can be made. The nature of the matters to be managed need to be considered. Whoever is appointed to manage the interface will be involved in a three-way process: they will be involved in clarifying accountabilities and responsibilities with the parties on either side of the interface; they will be involved in defining their own role; and they will be involved in liaising and working with those charged with the overall project interface management.

Care should be taken to identify the absolutely critical interfaces: these must be adequately resourced, even possibly over-resourced, which implies putting in contingencies to reduce their critical nature.

Using networked organisations

Some industries operate in what is coming to be identified as network organisations. For instance, in the Australian construction industry, quite a large proportion of subcontractors work with three or fewer contractors.[105] There is no formal, legal link between the contractor and subcontractor. This experience of working together, however, serves to increase the knowledge one has of the other's work methods, and thus reduces the interface problem. Some of these arrangements may run foul of competition laws. If this practice had to cease, it would be unfortunate—the project industry needs network-type organisations.

Meeting to co-ordinate interfaces

The major resource-consuming activities in projects devoted to interface management are meetings. Start-up meetings are a useful device to set good interface management in place.[106] In so far as interface management is concerned, the purpose of start-up meetings is to reach agreement on a number of significant operational details—to agree on who is responsible for what information, how it is to be moved around, who has authority for what service, and so on. If they are well-run, these meetings can provide a much-needed impetus to the project and set up positive relationships between participants.

In running the interface meeting, the project manager should concentrate on the co-ordination of the factors that are involved in the interfaces or that affect more than one participant.

The meetings also act to co-ordinate people. They bring together those project participants whose work affects other work. If a project participant's work does not affect the work of others, then it is probably unlikely that they need to attend a meeting. The project manager needs to group the various participants on the basis of their interaction and organise different meetings to suit each group. That the meetings are arranged like this will be appreciated by the project participants.

Where an interface problem involves only two project participants, it should be referred out of the meeting to be solved by those participants.

With a problem that initially only involves one participant but which will later affect others, the project manager needs to use the meeting to highlight the importance of dealing with it. Commitments should be obtained from the relevant participants to resolve the problem at and by a specific date. Peer pressure is an important device in this process.

There should be no surprises at the meeting. If people are required to make a decision, they should be briefed beforehand so that they can give the matter proper consideration. Springing surprises on people delays decision-making. (Springing surprises at meetings can also, of course, be a tactic to disadvantage opponents.)

Employing standardisation

This is crucial in interface management. It is an obvious advantage, though often difficult to achieve, if physical components actually fit together. Problems can arise, especially in the area of data flow, and some are practically impossible to solve during the life of the project. 'Quick and dirty' solutions are therefore needed. In software, this might mean writing a one-off conversion routine which will only last for the life of the project.

Standardisation is difficult to come to grips with. While it is clearly relevant, it is difficult to identify all the places or times where it is needed. With standardisation, one has to identify the need before the need is actually there, and by then it is too late. This is an area in which project management organisations can build up their stock of standards and history of the need for standards.

Standards are often the most boring documents to read—an exciting standard is almost impossible to imagine. However, it has been observed that while the words in a standard can be very boring, they can actually hide quite exciting political fights and compromises; the words tend to obscure the blood and guts in the background.

Having identified the areas needing standards, the project manager does not have to invent them. Many exist; the main problem is to get agreement on which ones are to apply.

Introducing people to each other

If people know each other, interface problems should be much alleviated (unless there have been earlier conflicts between them). Many problems can be quickly solved by a telephone conversation with someone familiar. If there is inherent conflict or a history of conflict between people, it is the project manager's business to know this in advance and to deal with it. Conflict only needs to be solved if it has the potential to adversely affect the project.

The project manager should make efforts to introduce interfacing participants to each other. These introductions can be formal or informal, whichever is most appropriate. The advantage of informal meetings is that information can be learned that is different from that which emerges at formal meetings. If formal processes are blocked, a few informal occasions such as lunch or a drink after work should be arranged.

As well as introducing individuals to each other, the interfacing organisations should be encouraged to set up mechanisms to maintain contact with each other.

Establishing processes for disputes resolution
Decisions at the interface affect people's success rates. They may also shift benefits from one side to another. Thus disputes are to be expected; negotiation skills are very relevant at the interface. Since individual negotiations may consume considerable time and energy, it might be appropriate to have a formal mechanism to arbitrate on disputes across the interfaces. Such mechanisms are not common, but on large projects with many interfaces some formal procedure might prove very useful.

Key points on interface management

The following serves as a check-list of some of the methods available to the project manager in the area of interface management. In implementing these methods, the project manager must be careful not to take on the responsibilities of the interfacing organisations but to only ensure that they actually take action to smooth the interfaces.

- identification of boundaries
- Work Breakdown Structures (WBS)
- interface matrices
- meetings
- monitoring progress
- configuration reviews
- reviews
- responsibility matrices
- communication
- use of common databases
- schedules
- boundary riders
- a special task force to check and approve work across boundaries
- liaison officers nominated by subcontractors
- people placed in other people's offices
- standardisation
- introduction of slack at critical points
- resource balancing
- planned access to critical resources
- unified terms and conditions of contract
- introductions between project participants
- disputes resolution process

Chapter Thirteen

13

And finally the Client

The client and the project manager have to work together to bring the project into existence. In carrying out their duties, each needs to know where the other stands and have reasonable expectations of how the other will perform. Although quite different, their roles complement each other. The client manages the project through the project manager; the project manager manages the project while keeping the client on-side and satisfied.

Management by the client

In some cases, the client decides to manage everything; however, here it is assumed that the client delegates the management of the project sequence to the client project manager. This does not completely free up the client, as there are project sequence decisions the client should make as well as work for which the client alone is responsible. The client must manage its own project manager, ensure the project manager is performing well in looking after the project, and make certain decisions. (Many of the comments here apply equally to the owner of the main supplier organisation, as the client for the main supplier project manager.)

Ensuring value for money

Perhaps the most important client responsibility is to ensure that money expended during the project sequence is spent wisely and obtains value in return. There needs to be a voice that keeps questioning expenditure and whether or not it is justified. The expenditure can only be fully justified and appreciated from a client perspective; it would be rare for a project manager to have sufficient access to the client's world to be able to see things exactly as the client would.

Making routine decisions

Projects require ongoing decision-making by the client. For any decisions to be taken, they need mechanisms through which the need for making a decision is identified, through which a decision is arrived at, and for the implementation of that decision.

Where the need for a decision arises directly from the project sequence, the client project manager should raise the issue and put forward a recommendation. Typical issues concern contracts to be signed and approvals to be given. Usually, the project manager is then charged with implementing the decision. As the project proceeds, the client will be asked to clarify certain requirements, to make decisions about variations and about trade-offs. Where such decisions involve considerable additional financial commitment, it is usual for the client itself to make the decision, although often with advice from the project manager. These can be quite difficult decisions to make, requiring considerable thought and analysis, and often having considerable impact on the project outcomes.

Monitoring the project

A client needs to assure itself that the project sequence is being managed properly. It also needs to protect itself against deceit or lack of full disclosure.

Things are going as reported

While actual deceit may not be easy to detect, its presence can be indicated. Time and cost management techniques are not intrinsically difficult to understand, and these same techniques should indicate any underlying problems. Keeping a close eye on the time and cost control procedures by means of regular inspections of work done can be quite effective (not detailed inspections, but ones that make possible a limited grasp of what is going on). The real danger is to assume that one cannot possibly understand what is going on and so not even take a look. Asking questions that seem to be based in ignorance can be a quite effective way of obtaining good, hard information—replying to simple, naive questions forces people to give simple answers; they have to come out from behind the security of jargon. Project managers are warned to treat people asking unexpectedly dumb questions with the greatest of care!

Bad news

On every project, there is bad news, and the client needs to know about it. The client may not be aware of the nature of the project sequence and the decisions that have to be made, but still needs to have a mechanism through which it continues to assure itself that all is proceeding as it should. A system should be in place whereby bad news is formally reported.

There is a strong culture against such reporting. There is a widespread view that only good news should travel. This is all very well if those hiding the bad news are committed and able to turn the bad news into good news, and there have been many cases of people concealing the bad news while they tried unsuccessfully to solve the problem (sometimes by illegal means). The point is that if the client does not know that the project is in danger, the client itself is in danger; it is also denied the opportunity to start taking action to avoid danger.

When things are not going well, there are signals, admittedly in a delayed form but signals all the same. These signals will be cost blowouts, time overruns, technical difficulties and, in some cases, new environmental data which challenge the assumptions underlying the project.

The primary responsibility for keeping the client informed lies with the client project manager. However, there are conflicts of interest here—reporting one's own bad news is not seen as a good career move. The client therefore needs to find an independent source or create a culture where delivering bad news is almost regarded as normal. The client should assume that project management means dealing with things going from bad to worse, with new threat after new threat, and endless unpleasant surprises. If the client is experiencing none of these, inquiry would be prudent.

Safeguards against dishonesty

The client needs to have independent sources of confirmation that people are acting honestly. In reality, this can be quite difficult. However, the client must put procedures in place that guard against a deceit that could jeopardise the project or the client itself. The most important part is to know where the money is coming from and where it is going. The client can use the services of lenders in determining where the money is coming from; this can done by people with accounting skills. Various industries have staff who specialise in the expenditure of money within that industry—in the construction industry, this work is often done by the quantity surveyor, who certifies that appropriate expenditure has taken place.

Managing the project manager

Clients must manage their own project managers—first they appoint them and then they manage their performance.

Appointment

The recruitment of a project manager can be quite difficult. It is hard to find people with all the necessary skills; even if they have the skills, it is difficult to *know* they have them! All the problems associated with executive selection are encountered.

However, the fact remains that somebody must be appointed. The selection of a project manager is a subjective process, depending on the social views of those making the selection. The client should have a procedure whereby the key requirements are identified and considered. Some of these requirements have been listed and described in chapter 1. The selection criteria should not, however, be totally inflexible when the final choice has to be made from the short-list of candidates. The selection process should require that, while the selection criteria should be given due weight, all relevant issues be considered. After consideration, there should be room for a best judgement.

An important issue to be examined in the selection process is that of values and attitudes. It is usually fairly easy to decide whether or not the candidate has the required technical skills, or whether or not the candidate could easily acquire them. What is not so easy to measure but which is equally important concerns the candidate's values and attitudes.

Performance review

The client does not wish to do, and should not do, the project manager's job. It is therefore recommended that the client nominates various points at which the client reviews the work of the project manager. These milestones can be simply a set of dates or they can relate to the project's progress—it depends on the level of confidence the client has in the project manager. The nature of the review depends on a range of factors, not unlike most other executive reviews.[107]

Authorising expenditure and defined limits

As long as the project is in control, the project manager or the person delegated to do so should be able to enter into the appropriate subcontracts and authorise expenditure. There is a technique using budget breakups which can be used to define the authority of the project manager. This overcomes the system of delegations, whereby authority to spend is limited by absolute amounts. For example, a delegation of a million dollars would be regarded as a very senior delegation in a large corporation, but the signing of a $1.2 million subcontract would be regarded as nothing out of the ordinary, particularly if the project budget exceeded $10 million.

A rule can be framed that states that someone may authorise an expenditure so long as it is below a certain percentage of the budgeted amount and that where there is any excess above the budgeted amount, that excess can be supplied from contingency, then that person may authorise the expenditure.

With such a rule, delegations can be determined by the extent to which the budget will be broken up. The more detailed the breakup, the lower the delegation. There can also be a hierarchy of delegations, as in the following example.

Train toilet: $110,000
Trade waste: $294,000
Stormwater: $215,000
Drainage: $145,000
Hydraulic services: $764,000

In this example, with a 10 per cent factor on the budgeted amount, the person with the authority over the hydraulic services has a delegation of $840,400. Thus that person can enter into a subcontract for hydraulic services as long as the value of the subcontract is below $764,000 plus $76,400 (ie, $840,400) and the contingency can supply the extra $76,400. Those looking after trade waste have a delegation of $323,400, if the hydraulic services are to be broken up into smaller subcontracts and each one let separately.

Most large organisations have delegations that are quite small in relation to required project expenditures. Having the chief executive involved in every project subcontract where that subcontract is under control is a waste of resources. This system of delegation can facilitate project administration.

Note that the question is raised as to whether the contingency has sufficient funds. The presence of an amount of money equal to or greater than the required amount is not necessarily sufficient evidence that the contingency has sufficient funds. The project manager must determine the appropriate funds to be available at a particular phase in the project. Thus at the end of the feasibility phase, the required contingency percentage is higher than at the end of planning and design.

The required contingency fund usually gets smaller as the project progresses. Careful monitoring of it provides vital data on the state of the project's budget. Where there is industry experience, an opinion has already been formed as to the appropriate level of contingency for each project phase. For example, a contingency of 30 per cent at the beginning of feasibility, 20 per cent at the beginning of planning and design and 10 per cent at the beginning of implementation might be seen as an industry norm. On this basis, it can be ascertained by examining the amount of contingency whether or not it is adequate and thus whether the budget is adequate.

This issue of an appropriate amount of contingency has even more relevance in fast-track projects where some parts are still in the early design stages while others are far advanced. Recognition should be given to the fact that parts are at different stages and the appropriate percentage should be applied to the amount at each stage, thus giving an estimate of the appropriate contingency.

Stating budget objectives in dollar values

There are many ways of stating the overall budget objective in dollar terms. One is the 'actual dollars spent' value and another is the dollar value as at a particular date. Stating the objective in terms of dollars spent is recommended for the following reason.

The present-day value statement permits an unnecessary relinquishing of control. While it is true that the project team does not have control over the market, it still can have control over its response. The project team may be able to change materials in response to a market price change—the fixed-date dollar value masks this possibility.

Taking corrective action

In order to protect its interests and the interests of the project, the client may have to take corrective action—in other words, directly interfere in the ongoing management of the project. This may include replacing the project manager. There are other forms of interference, such as directing that certain items be examined, cutting the available budget, abandoning the project, putting the project on hold (even in the middle of implementation) until some problems are fixed, and altering the scope of the project.

Such dramatic and powerful actions can be very successful. The author has seen a project stopped mid-implementation because of uncertainty about the budget; after a new budget was agreed, the project proceeded. This certainly put discipline into the cost control and scope management sequences.

Maintaining support for the project

Both the client and the client project manager must constantly work to maintain support for the project. Frequently, each has different forms of access to different people and different power groups. The client needs to ensure that the client project manager is working to obtain and maintain support, while at the same time gathering and husbanding its own sources of support.

The client's role is vital

The client's role is crucial for the success of the project. The client gives a sense of direction to the work of the project sequence, and provides a set of attitudes and values for the project management and a discipline to the expenditure of resources. This role requires commitment and time. Many clients abandon it on the assumption it is an impossible role and that everything should be left to the experts.

This text has tried to demonstrate that the experts can be managed and that considerable control over the project is possible—but it is only possible with the active support of the client.

Notes

1. P. W. G. Morris, *The Management of Projects*, Thomas Telford, London, 1994, provides a detailed history of the ideas behind project management.

Other views on the history of project management are R. D. Archibald, 'The history of modern project management: key milestones in the early PERT/CPM/PDM days', *Project Management Journal*, 18, no. 4, 29–31 Sept. 1987.

Alan Stretton, 'A short history of modern project management', Parts 1, 2 & 3, *The Australian Project Manager*, March, July and October 1994.

2. For a sample discussion on the usefulness of manuals, see Arne H. Omsen & Ollle von Post, 'Managing product development projects by company manuals', IPMA 1996 World Congress on Project Management and 12th AFITEP Annual Meeting, 24–26 June 1996, CNIT, Paris La Defense, France.

3. For descriptions of the various techniques, the reader is referred to other project management texts, such as Gesellshaft fur Projektmanagement INTERNET Deutschland e.V. *Projektmanagement Fachmann* Rationalisierungs-Kuratorium der Deutschen Wirtschaft e.V.,1991.

F. L. Harrison, *Advanced Project Management—A Structured Approach*, 3rd edn, Gower, 1992.

J. R. Meredith & S. J. Mantel, *Project management: a managerial approach*, 2nd edn, Riley, New York, 1989.

J. J. Moder, C. R. Phillips & E. W. Davis, *Project Management with CPM, PERT, and precedence diagramming*, 3rd edn, Van Nostrand Reinhold, New York, 1983.

John M. Nicholas, *Managing business and engineering projects: concepts and implementation*, Prentice Hall Inc, 1990.

M. Spinner, *Elements of project management: plan, schedule, and control*, 2nd edn, Prentice Hall, 1992.

4. Morris, 1994 (note 1).

5. Gesellshaft fur Projektmanagement INTERNET Deutschland e.V. *Projektmanagement Fachmann* Rationalisierungs-Kuratorium der Deutschen Wirtschaft e.V.,1991, p. 160.

6. R. Youker, 'A new look at the WBS: Project Breakdown Structure (PBS)', *PMNETwork*, vol. 5, no. 8, November 1991, Project Management Institute, pp. 33–6, 59.

Rene W. Hanssen & B. W. Schellekens, 'Setting up work breakdown structures for large-scale engineering projects', IPMA 1996 World

Congress on Project Management and 12th AFITEP Annual Meeting, 24–26 June 1996, CNIT, Paris La Defense, France.

7. See various US armed forces and US Department of Defense publications such as US Department of Defense, *Cost/Schedule Control Systems Criteria: joint implementation procedures*, Washington DC Government Printing Office, 26 August 1970.
For a quick introduction, see Lee R. Lambert, 'Cost/schedule control systems criteria (C/SCSC): an integrated project management approach using earned value techniques' in *AMA Handbook of Project Management*, ed. Paul C. Dinsmore, AMA Publications, New York, 1993, chapter 13, pp. 177–96.
Lee R. Lambert, *The cost/schedule control systems criteria (C/SCSC): an integrated project management approach using earned value techniques: a light-hearted overview and quick reference user's guide*, Lee Lambert & Associates, 1990.
Quentin W. Fleming, *Cost/Schedule Control Systems Criteria: the management guide to C/SCSC*, Probus, 1992.

8. For descriptions, see J. J. Moder, C. R. Phillips & E. W. Davis, *Project Management with CPM, PERT, and precedence diagramming*, 3rd edn, Van Nostrand Reinhold, New York, 1983.

9. For an excellent description of configuration management, see E. H. Bersoff, V. D. Henderson, S. C. Siegel, *Software Configuration Management—An investment in product integrity*, Prentice Hall, 1980. One can also refer to the International Standards Organisation Standard, ISO 1007: 1994, 'Guidelines for Configfuration Management'.

10. Some texts on risk are P. A. Thompson & J. G. Perry (eds), *Engineering Construction Risks—A guide to project risk analysis and assessment implications for project clients and project managers*, Thomas Telford, London, 1992.
C. Perrow, *Normal accidents: living with high-risk technologies*, Basic Books, New York, 1984.
Vlasta Molak, *Fundamentals of risk analysis and risk management*, CRC Press Inc., 1997.
Some recent papers on risk in the project management area are J. A. Bowers, 'Data for Project Risk Analyses', *International Journal of Project Management*, vol.12, no.1, 1994, pp. 9–16.
A. Oldfield & M. Ocock, 'Project Risk Management: The Importance of Stakeholders', *Proceedings of 12th INTERNET World Congress on Project Management*, Oslo, 1994.
P. Thompson & C. Norris, 'The Perception, Analysis and Management of Financial Risk in Engineering Projects', *Proc. ICE*, vol. 97, issue 1, Feb. 1993.

R. W. Stewart & J. Fortune, 'Application of systems thinking to the identification, avoidance and prevention of risk', *International Journal of Project Management*, vol. 13, no. 5, Oct. 1995, pp. 279–86.
J. Tuman, 'The Psychology of Choice in Project Execution Decision-Making and Risk Management', *Proceedings of 12th INTERNET World Congress on Project Management*, Oslo, 1994.

11. Peter Thompson, 'The contribution of teamwork to project risk management', IPMA 1996 World Congress on Project Management and 12th AFITEP Annual Meeting, 24–26 June, 1996, CNIT, Paris La Defense, Paris.

12. Professor Alan M. Stretton's work has greatly helped the author in this area.

13. J. W. Lorsch & P. R. Lawrence (eds), *Studies in organisational design*, Irwin-Dorsey, Homewood, Il., 1970.
P. R. Lawrence & J. W. Lorsch, *Organisation and environment: managing differentiation and integration*, Harvard University Press, Boston, 1967.

14. A. V. Feigenbaum, *Total quality management*, 3rd edn, McGraw-Hill, New York, 1983.

15. Sven R. Hed, *Project Control Manual*, published by Sven R. Hed, Case postale 330, CH-1211, Geneva 3, Switzerland, p. 43. This manual separates the project fron the project sequence but uses different terminology.

16. A. M. Stretton, 'Distinctive features of project management: a comparison with conventional management', *General Engineering Transactions*, I.E. Aust., vol. GE9, no.1, July 1983, pp. 15–21; this is an excellent paper teasing out the nature of projects and project management.

17. The idea of 'continual elaboration' turns up in Francis M. Webster, 'What project management is all about', *AMA Handbook of Project Management*, ed. Paul C. Dinsmore, AMA Publications, New York, 1993, p.11. Close approximations turn up in other texts, such as 'progressive definization' in Thomas T. Samaras & Frank L. Czerwinski, *Fundamentals of configuration management*, John Wiley & Sons, 1971, p. 3.

18. In some cases, the discovery of the sequence may be a major problem. See J. R. Turner & R. A. Cochrane, 'Goals and methods matrix: coping with projects with ill-defined goals and/or methods of achieving them', *International Journal of Project Management*, vol. 11, no. 2, 1993, pp. 93–102.

Notes

19. The importance of the social and political environment is highlighted in Morris, 1994 (note 1), and in P. W. G. Morris & G. H. Hough, *The anatomy of major projects*, John Wiley & Sons, Chichester, 1988. The importance is further highlighted by the apparent success and then demise of the Super Conducting Supercollider project. This project is the showcase project in the November 1990 edition of *pmNETwork* and its demise is reported in Willard E. Payson, 'The Superconducting Super Collider—the demise of a super project', *Proceedings of the 12th INTERNET World Congress on Project Management*, Oslo, 1994.

20. The reader is referred to the general literature on technological change and its impact.

21. The reader is referred to the general literature on organisation theory such as Fremont E. Kast & James E. Rosenzweig, *Organisation and management: a systems and contingency approach*, 4th edn, McGraw-Hill, 1985.

22. D. I. Cleland (ed.), *Matrix management systems handbook*, Van Nostrand Reinhold, New York, 1984.
Examples of discussions on replacing the functional hierarchy with matrix or project organisation can be found in the following papers presented at the IPMA 1996 World Congress on Project Management & 12th AFITEP Annual Meeting 24–26 June, 1996, CNIT, Paris La Defense, France: Raymond Savoye, 'Acceleration in the development of new vehicles with line/project teams towards an organization with double logic'; Zohra Alouani, 'Behaviour and relational aspects in the management of a concurrent engineering project: a sociotechnical approach'; and Randy Englund, 'Project and process strategies for future success'.

23. For discussions on structure, technology and function, see J. Woodward, *Industrial organisation: theory and practice*, 2nd edn, Oxford University Press, London, 1980; Lorsch & Lawrence, 1970.

24. This list of responsibilities owes much to the excellent book by Alan Webb, *Managing Innovative Projects*, Chapman & Hall, London, 1994, eg, on p. 128.

25. Webb, p.128 (note 24).

26. Robert Block, *The politics of projects*, Yourdon Press, a Prentice Hall Company, 1983.
Deborah S. Kezsbom, Donald L. Schilling & Katherine A. Edward, *Dynamic project management: a practical guide for managers and engineers*, John Wiley & Sons, 1989, p. 184.

Robert B. Youker, 'Power and politics in project management', *Proceedings of 10th INTERNET World Congress on Project Management*, Vienna, 1990.

27. For instance, see Morris & Hough, p. 258 (note 19).

28. Kurt Lewin, 'Field theory in social science', D. Cartwright (ed.), Tavistock, London, 1952, cited in Stanley B. Petzall, Christopher T. Selvarajah & Quentin F. Willis, *Management: a behavioural approach*, Longman Cheshire, 1991.

29. *ISMI Project Management 1*, The Australian Information Systems Management Institute.

30. *ISMI Project Management 1*, The Australian Information Systems Management Institute. See also William W. Agresti (ed.), *New paradigms for software development*, IEEE Computer Society Press, North Holland, 1986, for various papers on the forms of the project life cycle.

31. The project advocate also might be called a 'product champion' or 'project champion'.

32. Reo M. Christenson, Alan S. Engek, Dan N. Jacobs, Mostafa Rejai & Herbert Waltzer, *Ideologies and Modern Politics*, 2nd edn, Harper & Row, 1975, p. 17.

33. Christenson, Engek, Jacobs, Rejai & Waltzer, p. 17 (note 32).

34. I. B. Myers, & M. H. McCaulley, *Manual: a guide to the development and use of the Myers-Briggs Type Indicator*, Consulting Psychologists Press, 1985.
I. B. Myers, *Gifts differing*, Consulting Psychologists Press, 1980.
G. Lawrence, *People types and tiger stripes: a practical guide to learning styles*, 2nd edn, Centre for Applications of Psychological Type, 1982.

35. M. Brassard, The Memory Jogger +™, GOAL/QBC, 13 Branch St., Methuen, Ma. 01844, 1989.

36. Morris, 1994, p. 218–221 (note 1).

37. D. Boddy & David A. Buchanan, *Take the lead: interpersonal skills for project managers*, Prentice Hall, New York, 1992.

38. Kaoru Ishikawa, *Guide to Quality Control*, 2nd edn, Asian Productivity Organisation, 1976, cited in W. E. Deming, *Out of crisis*, Massachusetts Institute of Technology, Cambridge, Mass., 1991.

39. Michel Thiry, 'Added value project management', IPMA 1996 World Congress on Project Management & 12th AFITEP Annual Meeting 24–26 June, 1996, CNIT, Paris La Defense, France
Alfred I. Paley, 'Value engineering and project management: achieving cost optimisation' in *AMA Handbook of Project Management*, ed Paul C. Dinsmore, AMA Publications, New York, 1993, chapter 14, pp. 197–204.

40. Mette Amtoff, 'Projects as a construction of many stories', *Proceedings of 12th INTERNET World Congress on Project Management*, Oslo, 1994.

41. Malcolm Hollick, *An introduction to project evaluation*, Longman Cheshire, Melbourne, 1993.

42. Thompson & Perry 1992; Perrow 1984; Vlasta 1997; and papers at note 10.

43. See publications at note 3.

44. ISO Standards Committee, 'Quality management: Guidelines to quality in project management' *(ISO/CD 10006 Committee Draft Document :176-2-8-N252)*, British Standards Institute, March 1995.

45. Amanda Gome, 'Total quality madness', *Business Review Weekly*, 30 Sept. 1996.

46. Rensis Likert, *The Human Organisation*, McGraw-Hill, 1967; Rensis Likert, *New patterns of management*, McGraw-Hill, 1961.

47. Russell D. Archibald, *Managing high-technology programs and projects*, 2nd edn, John Wiley & Sons Inc., 1992, chapter 11.
M. Fangel, *Project start-up manual*, Internet, Zurich, 1987.
O. Husby & S. Skogen, 'A new systematic approach to the general start-up challenges', *Proceedings of 12th INTERNET World Congress on Project Management*, Oslo,1994.

48. Lecture by Mr Tom Pettigrew, Product Quality Assurance Manager for Ford Australia, to Master of Project Management program at the University of Technology, Sydney, 1989.

49. 'House of quality' refers to 'Quality Function Deployment'.
James L. Bossert, *Quality Function Deployment: a practitioner's approach*, ASQC Quality Press, Milwaukee, Wis, 1991.
N. Logothetis, *Managing for total quality: from Deming to Taguchi and SPC*, Prentice Hall, 1992.
Yoji Akao (ed.), *Quality function deployment: QFD integrating customer requirements into product design*, Productivity Press, 1990.

50. Brassard 1989 (note 35).

51. Brassard 1989 (note 35).

52. Lecture by Mr Tom Pettigrew, Product Quality Assurance Manager for Ford Australia, University of Technology, Sydney, 1989. Pettigrew has written a number of articles on quality, among which are: T. J. Pettigrew, 'Process quality control: the new approach in the management of quality at Ford', *Journal of Society of Automotive Engineers Australia*, July/August 1983; 'Quality, cost efficiency and integrity in the manufacture of goods: Australia can match the world's best', Annual Engineering Conference, I. E. Aust., Melbourne, Australia, 2–8 March 1985.

53. Feigenbaum, 1983 (note 14).

54. Igor S. Popovich, *Managing Consultants: How to choose and work with consultants*, Century Ltd., Random House, 1995.

55. Archibald, 1992, chapter 11; Fangel, 1987; Husby & Skogen, 1994 (note 47).

56. Boddy & Buchanan, 1992 (note 37).

57. Archibald, 1992, chapter 11; Fangel, 1987; Husby & Skogen, 1994 (note 47).

58. Amtoff, 1994 (note 40).

59. Webb, 1994 (note 24), provides a description of the slip diagram, pp.189–92.

60. Morris, 1994 (note 1), p. 222 and p. 328 (the Skylab falling example).

61. Ishikawa, 1976, cited in Deming, 1991 (note 38).

62. Ishikawa, 1976, cited in Deming, 1991 (note 38).

63. Systems Gap Working Party Report, Association of Project Managers, *Closing the gaps in project management systems*, Butterworths, 1984.

64. Endre Y. Mozes, 'Teams and Techniques in Schedule Compression for Mega-Projects', IPMA 1996 World Congress on Project Management & 12th AFITEP Annual Meeting, 24–26 June 1996, CNIT, Paris La Defense, France.

65. R. Fisher & W. Ury, *Getting to YES: Negotiating Agreement Without Giving In*, Houghton Mifflin, 1981.

66. David A. Hamburger & Spirer F. Herbert, 'Project Completion', pp. 587–616, in Robert L. Kimmons & James H. Loweree, *Project Management: A reference for professionals*, Marcel Dekker, 1989, is an excellent reference on the management of the handover phase of technical projects.

67. Systems Gap Working Party Report, Association of Project Managers, 1984.

68. Archibald, 1992, chapter 11; Fangel, 1987; Husby & Skogen, 1994 (note 47).

69. The concept of the downstream customer, which has come from the quality movement, is a useful idea that can be applied in the project sequence.

70. Boddy & Buchanan, 1992 (note 37).

71. Video Arts, *So you think you can sell*, Methuen, 1985.
Malcolm H. B. McDonald & John W. Leppard, *How to sell a service: guidelines for effective selling in a service business*, 2nd edn, Butterworth Heinemann, 1988.

72. Herb Cohen, *You can negotiate anything*, Angus & Robertson Publishers, 1980.
Gavin Kennedy, *Everything is negotiable*, Business Books, 1982.
Margaret A. Neale & Max H. Bazerman, *Cognition and rationality in negotiation*, The Free Press, 1991.

73. Morris, 1994 (note 1), gives a detailed description of the concurrency debate.

74. Deming, 1991 (note 38); Feigenbaum, 1983 (note 14); J. M. Juran, *Juran on Leadership for Quality: An Executive Handbook*, The Free Press, a division of Macmillan, Inc.
Papers presented at IPMA 1996 World Congress on Project Management & 12th AFITEP Annual Meeting, 24–26 June 1996, CNIT, Paris La Defense, France: W. Hagen, 'Quality management in IBM Switzerland, an experience'; Jorg-Martin Hohberg, 'Quality management in construction projects—rivaling project management?

75. Gome, 1996 (note 45).

76. See publications at note 10.

77. See publications at note 74.

78. David M. Himmelblau, *Process Analysis by Statistical Methods*, John Wiley & Sons, Inc., 1970.

79. T-Group training has been studied widely and is not without its critics. For a summary of the issue, see Petzall, Selvarajah & Willis, 1991, p. 251–2.

80. Richard Bandler & John Grinder, *The structure of magic*, Science & Behaviour Books, 1975.
Richard Bandler & John Grinder, *Reframing: neuro-linguistic programming and the transformation of meaning*, Real People Books, 1982.
Phillip J. Boas & Jane Brooks, *How to make a frog into a prince(ss): an NLP training manual*, Magic Mouse, 1984.

81. Video Arts, 1985; McDonald & Leppard, 1988 (note 71).

82. Cohen, 1980; Kennedy, 1982; Neale & Bazerman, 1991 (note 72).

83. Fisher & Ury, 1981 (note 65).

84. Bandler & Grinder, 1975; Bandler & Grinder, 1982; Boas & Brooks, 1984 (note 80).

85. Fisher & Ury, 1981 (note 65).

86. Cohen, 1980; Kennedy, 1982; Neale & Bazerman, 1991 (note 72).

87. Video Arts, 1985; McDonald & Leppard, 1988 (note 71).

88. Hans J. Thamhain, 'Effective leadership for building project teams, motivating people, and creating optimal organisational structures' in *AMA Handbook of Project Management*, ed. Paul C. Dinsmore, AMA Publications, New York,1993, chapter 19.
Nicki S. Kirchof & John R. Adams, *Conflict management for project managers*, Project Management Institute, 1982.

89. The study of group behaviour is a fascinating area with an extensive literature. For a useful summary, one might refer to Petzall, Selvarajah & Willis, 1991; and Kast & Rosenzweig, 1985.

90. Petzall, Selvarajah & Willis, 1991 (note 79).

91. Petzall, Selvarajah & Willis, 1991 (note 79), provides a brief summary of group development.

92. R. M. Belbin, *Management Teams: why they succeed or fail*, Butterworth Heinemann, 1981;
R. M. Belbin, *Team roles at work*, Butterworth Heinemann, 1993.

93. Petzall, Selvarajah & Willis, 1991 (note 79).

94. Belbin, 1981; Belbin, 1993 (note 92).

95. Amtoff, 1994 (note 40).

96. Pierre Ryckmans, 'The view from the bridge: Aspects of culture', *The 1996 Boyer Lectures*, ABC Books 1996, p. 20, quoting from Zhuang Zi, *In the world of men*, chapter 4.

97. Fred Fiedler, *A theory of leadership effectiveness*, McGraw-Hill,1967.

98. Morris, 1994 (note 1); Morris & Hough,1988 (note 19), provide excellent evidence and overview of the importance of the external environment.

99. Reo M. Christenson, Alan S. Engek, Dan N. Jacobs, Mostafa Rejai & Herbert Waltzer, *Ideologies and Modern Politics*, 2nd edn, Harper & Row, 1975, discusses a definition of ideology.

100. Amtoff, 1994 (note 40).

101. Alan Stretton, 'A note on project scope', *The Australian Project Manager*, vol.14, no. 5, March 1995.

102. Bersoff, Henderson & Siegel 1980 (note 9).

103. Lewin, 1952, cited in Petzall, Selvarajah & Willis, 1991 (note 28).

104. Archibald, 1992 (note 47), discusses interface management in chapter 13.
P. W. G. Morris, 'Managing project interfaces—key points for project success' in *The Project Management Handbook*, eds D. I. Cleland & W. R. King, 2nd edn, Van Nostrand Reinhold, 1988.

105. Royal Commission into Productivity in The Building Industry in New South Wales, 1992.

106. Archibald, 1992, chapter 11; Fangel, 1987; Husby & Skogen, 1994 (note 47).

107. P. Zeman & P. Healy, 'Dynamic Leadership: The Role of the Client', *Proceedings of 12th INTERNET World Congress on Project Management*, Oslo, 1994.

Further Reading

Burrill, C.W. & Ellsworth, L.W. *Modern project management: foundations for quality and productivity*. Burrill-Ellsworth Associates, 1980.

Cleland, D. I. & King, W. R. (eds). *The Project Management Handbook*. 2nd edn, Van Nostrand Reinhold, 1988.

Dinsmore, Paul C. (ed.). *AMA Handbook of Project Management*. AMA Publications, New York, 1993.

Gareis, R. (ed.). *The Handbook of Management by Projects*. Manz, Vienna, 1990.

Harvard Business Review. 'Managing projects and programs' (ISBN 0-86735-106-3).

Kerzner, H. *Project management: a systems approach to planning, scheduling and controlling*. 5th edn, Van Nostrand Reinhold, New York, 1995.

Lock, Dennis. *Project management*. 4th edn, Gower, 1988.

Metzger, Philip W. & Boddie, John. *Managing a programming project, people and processes*. 3rd edn, Prentice Hall, 1996.

Silvermann, Melvin. *The art of managing technical projects*. Prentice Hall, 1987.

Turner, J. Rodney. *Handbook of project-based management: improving the processes for achieving strategic objectives*. McGraw-Hill Book Company Europe, 1993.

Index

A
acceptance tests, 198–200
accounting system, 187
authority, delegated, 27–9, 282–3

B
bias, 85, 89
bids, 98–9
boundary management, 140
breakdown structures, 3
budget control, 27, 49, 137–9, 182–3, 282–4

C
client, 59–60
  approvals, 140, 160
  functions of, 171, 173, 174, 175–80, 188–90
  management by, 279–84
client project manager, *see* project manager
commissioning, 194, 203–5
concept, 49–51,117–22
concurrency, 215–6
configuration management, 4, 257
consultants, 130–3, 140–1
contracts, 100, 107–9, 139, 183, 196–7
contractual relationships, 28–9, 107–9
control, 22–3, 27–9, 160
  of core technology, 265–6
  of project sequence, 151–6
control framework, 41–52
  and cost control, 48–51
  and phases, 45–7
  and scope management, 259–61
  and time control, 51–2
controlling cliques, 65–6
co-ordination, 25–7
core technology, 265–6
cost-benefit analysis, 89–90
cost contingency, 81, 137–9, 283
cost control, 48–51, 137–9, 182–3, 282–4
cost estimating, 80–2, 139, 220–1
crashing, 143–4, 182

D
design, 75–7, 101–2, 140–4
design and supply, 95–6
disclosure, 247–8
disputes resolution, 205, 278
documentation, 51, 83–4, 117–8, 122–5, 144–5, 188, 197
downstream customer, 212–3

E
earned value techniques, 4
environment, 79, 91, 100
equipment, specialised, 102–3
expenditure authority, 27, 282–3

F
fast-track, 36, 40–1
feasibility phase, 49, chapter 4
  brief for, 71
  objectives and scope, 260
finance, 90–1, 176–7, 183

G
government approvals, 160
group dynamics, 236–9

H
handover, chapter 8, 212–3

I
ideology, 244–6
implementation phases, 51, 146–9, 153–5, chapter 7
industrial relations, 101, 160, 186–7
interfaces, chapter 12

J
jargon, 264–5
juries, 151

L
lead times, 159, 163
leadership, 239–42
legal issues, 91, 101, 107–9, 186, 202, 248
logistics, 101

M

make or buy, 94–5
materials handling, 185
meetings, 134–6, 191–3, 276–8
morale, 178, 188

N

needs, defining, 77
negotiation, 229–30
network techniques, 4
networked organisation, 234–6, 276
neuro-linguistic programming, 228, 229
non-performance, 184
norms, 237
novation, 96

O

Organisation Breakdown Structure, 3

P

parallel feasibility phase, 208–9
partnering, 255
payments, 51, 177–8, 183, 184, 202–3
planning and design phase, chapter 5
  concept, 118–22
  documentation, 122–5
  objectives and scope, 136, 260–1
  organisation, 128–30
  relationships, 132–4
  risks, 141–4
  staff, 130–3, 140–1
political environment, 242–4, 246–9
political interests, 64, 88, 187, 213–4
practical completion, 195
pre-implementation phase, 51, chapter 6
process management, 10, 221–3
Product Breakdown Structure, 3
Product of Implementation Phase, 201–2
production infrastructure, 148–9, 150–1, 156–8
project
  acquisition, chapter 2, 93
  capability, 207
  champion, 215
  clarification, 74–84, 178
  client, *see* client
  communication, 191, chapter 10
  completion, 12, 14, 195
  definition of, 9, 12–13
  effectiveness, 70–1
  elaboration of, 13, 171, 178
  extent of, 77–80
  infrastructure, 103–4, 148–9
  manuals, 160, 163–4
  measurements, 149–51
  opposition to, 66
  ownership, 57–9, 188–9
  and project sequence, 12–13
  proposal, 83–93, 136
  selling of, 230–2
  social component of, 78–80, 91
  specifications, 139, 198
  stakeholders, 61, 67
  story, 87, 114, 251
  support for, 63–8, 189–90, 213–5, 242–4, 284
  teams, 105–6
  users, 60–1, 190
project advocate, 56, 57, 58–9, 64, 65, 66, 67, 68–71, 110
project control group, 62–3, 191
project life cycle, 31–41, *see also* control framework
project management
  characteristics of, 13–16
  definition of, 9, 10, 13
  as a discipline, 1–3
  generic, 4–5
  information systems, 144–5, 163, 218–20
  organisational forms, 22–30
  skills needed, 16–19
  social impact of, 19–30
  techniques of, 3–4, 10–12
project manager
  client, 61–2, 110–1, 173, 174, 175–80, 188–90, 281–2
  main supplier, 173–4, 180–8, 190
project organisation, 104–7, 164–5, 234–6, 276

project phases, 32–6, 39–41, 45–7, 49–51, part two
project sequence
 to achieve project, 93–103
 capability, chapter 9
 controllable aspects of, 151–6
 definition of, 9, 12–13
 financing, 183
 and industrial relations, 186–7
 management of, 16–19, 37, 109–10
 measuring in, 149–51
protest groups, 246, 248–9
prototyping, 103, 122
public relations, 189, 242, 243–4, 247

Q

quality management, 100–1, 125–8, 172, 216–7, 222–3

R

resources, 14–15, 88–9, 189–90
responsibility matrix, 212
retentions, 197
risk management, 4, 91–2, 141–4, 217–8
risky technology, 262–3
role theory, 238–9

S

safety, 101, 160, 186
scope management, 80, 252–61
standardisation, 277
start-up meetings, 113–4, 134–6, 175, 276
subcontractors, 15, 159, 167–8, 180–1, 184–5
supplier, main, 159, 166–8, 171–2, 173–4, 180–8, 190
suppliers, 97–8, 133–4
symbolic activity, 240–1

T

task responsibility matrix, 3
technology
 changes in, 19–20
 coping with, 15–16, 26, 263–5
 core, 265–6
 managing, 26–7, 78, 215–6, 261–6
 risks of, 141–2, 262–3
tender packages, 161–2
tendering, 96–7
time contingency, 38, 83, 143
time estimating, 82–3, 220–1
time management, 14–15, 38, 51–2, 92, 142–3, 181–2
trade-offs, 15, 136
transition phase, chapter 3, 259–60
turnkey approach, 95
two-boss situation, 29–30

V

variations, 171, 172, 179, 187, 255, 256–7

W

warranties, 197–8
Work Breakdown Structure, 3, 80, 94, 140
work packages, 161–2, 171–2, 180–1, 182–3